The Great Scientists
OF THE WORLD

Ashwani Bhardwaj

GOODWILL PUBLISHING HOUSE®
B-3 RATTAN JYOTI, 18 RAJENDRA PLACE
NEW DELHI -110008 (INDIA)

© Publishers

All rights reserved. No part of this publication may be reproduced; stored in a retrieval system or transmitted in any form or by any means, mechanical, photocopying or otherwise without the prior written permission of the publisher.

Published by
GOODWILL PUBLISHING HOUSE®
B-3 Rattan Jyoti, 18 Rajendra Place
New Delhi-110008 (INDIA)
Tel. : 25750801, 25820556
Fax : 91-11-25764396
E-mail : goodwillpub@vsnl.net
website : www.goodwillpublishinghouse.com

Printed at :-
Kumar Offset Printers
New Delhi

CONTENTS

1. Albert Einstein 1
2. Andreas Vesalius 8
3. Antonie Van Leeuwenhoek 12
4. Archimedes of Syracuse 16
5. Birbal Sahni 24
6. Charles Babbage 30
7. C. V. Raman, Sir 36
8. Charles Goodyear ... 39
9. Charles Parson 48
10. Charles Robert Darwin 52
11. Christiaan Barnard 64
12. Christopher Wren, Sir 67
13. Count Zeppelin 75
14. Don Bateman 80
15. Edward Jenner 88

16. Eli Whitney 96
17. Elias Howe 105
18. Enrico Fermi 113
19. Ernest Rutherford 114
20. Edmund Halley 122
21. Frank Whittle 126
22. Galileo Galilei 132
23. George Stephenson 140
24. Georges Brossard .. 152
25. Gregor Mendel 157
26. H.J. Bhabha, Dr. ... 161
27. Hargobind Khorana, Dr. 163
28. Henry Ford 165
29. Ian Wilmut, Dr Dolly 174
30. Isaac Newton, Sir .. 182
31. Jagdish Chandra Bose, Sir 189
32. J.J. Thomson, Sir .. 195
33. James Watt 198
34. James Young Simpson 207
35. John Baird 211

36. John Flamsteed 217
37. Joseph Baron Lister 224
38. Kamala Sohonie 234
39. Kolachala Seeta Ramayya, Dr. 238
40. Konstantin E. Tsiolkovsky 240
41. Leonardo Da Vinci 250
42. Linus Carl Pauling 261
43. Louis Pasteur 266
44. Meghnad Saha 282
45. Michael Faraday .. 286
46. Pierre and Marie Curie 297
47. Robert Goddard 305
48. Reuven Feuerstein, Dr. 311
49. Robert Watson-watt 318
50. Salim Ali 325

51. Samuel Hahnemann, Dr. ... 330
52. Satish Dhawan 333
53. Satyendra Nath Bose 334
54. Shanti Swarup Bhatnagar, Dr. 341
55. Srinivasa Ramanujan .... 344  
56. Thomas Alva Edison 350
57. Vikram Sarabhai, Dr. 357
58. Wilbur and Orville Wright 360
59. Wilhelm Roentgen 371
60. William Caxton .... 374 
61. William Harvey 382
62. William H. Perkins 389
63. William Herschel .. 395
64. William Shockley .. 398
65. William Thomson, Sir (Lord Kelvin) ... 406

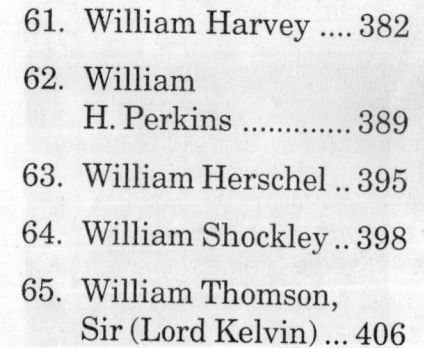

Albert Einstein

(1879–1955)

For the first week of November, 1919, England seemed to be settling down to observe the anniversary of the armistice. The bloodiest war in history had been over for a year, the world was back to normal, even if people felt things had changed: the cost of living, for example, the behaviour of young people (some critics even included the weather), all this might have changed for the worse, but there was much to give thanks for. The London Press rose to the occasion with special articles, editorials, sombre headlines. The momentous anniversary seemed to be occupying people's minds to the exclusion of all else.

November 7 brought a different headline. *The Times* printed, in huge letters, "Revolution in Science—Newtonian Ideas Overthrown" and went on to describe the previous evening's meeting of the Royal Society. In this, the results had been announced of the expedition to the Gulf of Guinea to view the solar eclipse. The observations of the expedition had been more than satisfactory: the paper quoted the Society's President:

"This is not the discovery of an outlying island but of a whole continent of new scientific ideas—it is the greatest discovery in connexion with gravitation since Newton enumerated his principles..."

In effect, the expedition's observations of stars during the period of the sun's total eclipse had verified the whole of Albert Einstein's theory of relativity, the astonishing theory of high-speed motion, the new, half-understood physics which seemed to unify all the branches of the subject. The old Newtonian theories were now proved inadequate when dealing with objects moving at tremendous speeds approaching that of light (186,000 miles a second), though they would remain accurate enough for the ordinary occurrences of life. As an example, Einstein had calculated that a clock, a timekeeping mechanism of whatever construction, which chanced to be travelling at 161,000 miles a second, would be registering time at exactly twice the rate of clock which was—relative to it—stationary. It would register that speed simply because time itself had speeded up. Not only time altered at these speeds; so did mass and dimension: at the same speed, a man travelling head first through space would be exactly half his usual length. (But as any yardstick he might carry to check this alarming fact would have shrunk by the same amount, he would be unaware of the change.)

These phenomena, and many more besides, had been arrived at by Einstein through a process of reasoning and mathematics. None of them apart from a peculiarity in the orbit of the planet Mercury, which his theories explained, had been observed or even suspected. There seemed little immediate chance of checking on the behaviour of objects at the speed of light, but it was generally conceded that if light could be observed to bend as it passed near the sun—a phenomenon predicted from the same calculation and one which most scientists considered ridiculous—then the whole theory could be taken as accurate. The 1919 eclipse of the sun provided just the opportunity. Two expeditions set sail for the latitude where the eclipse would be seen as total, one to the Gulf of Guinea, West Africa, the other to northern Brazil, each as it insurance against bad weather conditions which might affect the other. In this latitude the moon coming between earth and sun would, for a few minutes, completely hide the sun. At the same time the sun would be in the midst of a group of particularly bright stars and with its light temporarily blotted out, photographs could be taken of the stars as their light passed close to its surface. The apparent position of the stars could then be compared with pictures of the same stars taken by night in London, when they were far removed from the sun and therefore much less exposed to its gravity.

The pictures taken during the eclipse triumphantly proved Einstein's theory, the star-images on the photographic plates were deflected by exactly the amount he had calculated. There could no longer be any doubt about the theory of relativity—it worked. Yet, as the President of the Royal Society went on to say, "I have to confess that no one has yet succeeded in stating in clear language what the theory of Einstein really is."

No one could. A few scientists understood the consequences of this new, all-embracing theory within their own special fields: none could as yet grasp the meaning of the theory itself.

Yet a few tears later the theory was the standard, essential tool of every physicist, the words "Einstein" and "Relativity" were being bandied about, with greater or less understanding, over half the world. At least one reason lay in the remarkable character of Albert Einstein himself, a strange blend of kindness, intelligence and *naivete* which gave Relativity a publicity no other theory in physics had ever enjoyed.

He was born on 14 March, 1879, in the town of Ulm in Wurttemberg but left with his parents when he was a year old and moved to the large south German town of Munich. His father had started a small electro-chemical factory there with his brother, and the family settled into a cottage in the suburbs of the town while the father busied himself (not too successfully) with the commercial aspects of his business and uncle took charge of the factory. It was from this uncle, a remarkable man by all accounts, that Albert got his interest in scientific and mathematical processes. From his mother, who was musician, he inherited a deep love of music.

The family was Jewish and the boy was brought up in the faith, but as there was no convenient Jewish school, he was sent to a Catholic one and studied that faith as well. This peculiarity of upbringing may have accounted for the lifelong hatred of prejudice and bigotry which characterized him, the reluctance to become identified with any particular nation or group. He developed an interest in science, deplored "the time one wastes learning Latin and Greek", and became fascinated by mathematics, largely through his uncle, who delighted in working through algebraic problems with him in the evening, much as a more normal uncle might read a story out loud. "The animal we are hunting, he's hidden, and we'll call his 'x' for the moment. But we'll get him, all right . . ."

When the boy was fifteen his father lost money, closed the Munich factory and went south to try his luck in Milan. As it was taken for granted that every educated German had a "Gymnasium" diploma, Albert stayed behind at his school, the Luitbold Gymnasium. But somehow the thought of sunny Italy, not a bit like cold Munich, appealed so much that he left the school, despite his parents' protests, and joined them. He was just in time to find his father bankrupt, unable to support him further. Deprived, through his own impatience, of the advantages of a Gymnasium diploma, the boy looked for some technical institution that might train him for a scientific career. Eventually, after a second migration, to Switzerland, he found that the Polytechnic Institute in Zurich would have him.

The Polytechnic had an international reputation and students flocked to it from all over Europe. One of these was a young girl from Hungary, Mileva Maritsch, who seemed to share the young Einstein's passion for physics. They would work together in the laboratory long after other students had retired to their lodgings, they became inseparable and within weeks it became tacitly understood that, some day, they would marry.

On leaving the Polytechnic, Einstein discovered, to his dismay, that no one would offer him the sort of work he wanted; after much inquiry he found himself a post in the patent office in Berne. He had come to love the country so he renounced his German nationality and became a Swiss subject. To complete his readjustment, he married Mileva, only to find almost immediately that, outside the laboratory, they had nothing in common: the girl was an earnest, reserved, suspicious Slav; he was a vague, happy-go-lucky, Bohemian. Two sons were born to them in rapid succession and these, because both parents adored them, kept the marriage together.

Einstein's work at the patent office was well paid and interesting: it consisted of making an investigation of every invention and it required an ability to pick out basic ideas from descriptions which were usually inaccurate and badly written. It left plenty of time for other studies.

In 1995 Einstein published papers on the production and transformation of light and on the electro-dynamics of moving bodies. These caused a small stir in the academic world: how could such research come from an official in a patent office? Investigations were made and it was decided that the young man must be taken from his unsuitable post and given some junior professorship. Things moved slowly; four years later Albert Einstein was appointed to the University of Zurich.

Here, life was very different. The salary was no higher than what he had been getting but now he had to keep up a social standard, wear socks that matched and a well-pressed suit. His wife felt at home in the new atmosphere but they were suddenly very poor—clothes and furniture cost money. As Einstein was to remark later, recalling the Zurich days, "In my Relativity I set up a clock at every point in space— but in reality, I found it difficult to provide even one in my room."

In 1911 he was appointed to a better-paid professorship in Prague. Here, there was a large Jewish community, of which he became a member and for the first time, aware of its problems. The German attitude to his people was hardening, there were unpleasant incidents which in the light of history can be seen as tiny pointers to the nightmare which lay thirty years ahead for the Jews of Germany. Therefore when Einstein was offered a post in Switzerland, teaching at his old Polytechnic, he was glad to go— and Mileva, who had been miserable in Prague, made sure they left immediately.

Publication of the results of more of his research made Einstein an international figure. He worked for only two years at the "Poly" before he was made the sort of offer no scientist can refuse: membership of the Royal Prussian Academy of Science and directorship of the new Kaiser Wilhelm Institute in Berlin. This would give him a larger salary and much more time to devote to research. He accepted, but this time Mileva refused to go with him. She knew she would hate Berlin—and in any case they were both agreed that the marriage, at last had failed. She stayed behind with the two boys.

Fortunately for Einstein a distant cousin, Elsa Einstein, was living in Berlin. She was a young widow, and she took good care of him. He returned the compliment a year later, by marrying her.

He spent twenty years, from 1913 to 1933, in Berlin, working hard on his theory of Relativity, thinking it out stage by stage, overcoming its mathematical difficulties, gradually evolving the all-embracing theory which has affected the work ever since of all mathematicians and physicists. His simple-sounding formula, E equals mc^2, which proved the unsuspected equivalence of energy and mass has since been abundantly proved and is responsible, among other things, for the invention of the atomic bomb. The formula shows that a small mass can be converted into huge amount of energy, where E is energy, m is mass and c is the speed of light. It also gives the secret of the sun. If it were really burning, as people thought, it should have been consumed long ago; in the atomic reaction which Einstein postulated huge quantities of light and heat could continue to be liberated with the loss of only very small mass.

Despite his preoccupation with work, he was far from happy with the spirit of militarism which surged up during the war. He refused to sign the various manifestos which were put in front of him, alleging things like "German Culture and German Militarism are Indentical", he took little interest in the prosecution of the war, and it was only his Swiss citizenship that saved him from going to prison as a "traitor". After the war, when his theories were beginning to rock the world, there was haggling as to whose property he was. As he put it himself in a wry, very typical, letter to the Press,

"... *by an application of the theory of Relativity to the taste of the reader, today in Germany I am called a German man of science; in England I am represented as a Swiss Jew. If I come to be regarded as a* **bete noire**, *the description will be reversed and I shall become a Swiss Jew for the Germans, a German for the English* ..."

By 1922 Einstein's own deep conviction that science is a religion and philosophy was becoming more general. In that year he was appointed to the League of Nations Commission for Intellectual Co-operation. He resigned a year later because he felt the League had "neither the strength nor the good will to accomplish its task".

Slowly he bean to realize that Germany, much as he enjoyed his work there, was no place for a Jew. Reluctantly, in 1933, he accepted one of the many offers which were coming his way and sailed with his wife to take up post at the new Institute of Advanced Study in Princeton, New Jersey.

To his great sorrow, Elsa, who had been unable to make the adjustment to the new world from the Germany she loved, died in 1936. He stayed on in Princeton. a shambling, good-natured genius, prepared to talk to anyone at any time. ("Why should I mind that you are late meeting me? Am I less capable of reflecting on my problems here than at home? Here ...", showing a much-chewed pencil, "here is my laboratory.")

He died in Princeton in 1955, *universally mourned as the world's and the century's, greatest man of science.*

Andreas Vesalius

(1514–1564)

The Beginning of Modern Anatomy

Much of our basic knowledge of human anatomy comes from the anatomists of the 16th and 17th centuries, who metriculously cut up corpses and then accurately drew what they saw.

Nowadays human bodies are dissected every day by students to learn anatomy. But it was not always so. The Law did not permit dissection. It was also impossible to get human bodies for dissection and some early anatomists even had to steal them from graves.

Andreas Vesalius was the surgeon who dissected the human body to make the first major study of its anatomy. Fourteen centuries earlier, the Greek surgeon Galen who lived in Rome had written a text on human anatomy. But he did so by dissecting the bodies of animals and not of humans. Therefore many errors had crept into his study. Nevertheless Galen who was the personal physician to the Roman Emperor Marcus Aurelius was a great medical scientist of the ancient world. His views dominated European medicine for over 1400 years till Vesalius challenged them.

Vesalius was born in Brussels, Belgium, in the year 1514. His father was pharmacist to the king. His grandfather and great grandfather were physicians. Perhaps because of this family tradition, Andreas had a liking towards medical studies from very early in life.

As a child he used to practice dissection on dogs, cats and other animals on his mother's kitchen table. At the age of sixteen he joined the University of Louvain in Belgium for studying medicine. He spent nearly three years there and then moved to the University of Paris in 1533, when he was nineteen.

In Paris, Vesalius studied under Jacob Sylvius, who was then the most important figure at the Faculty of Medicine in the University. He earned a good reputation as an intelligent and hardworking student, but couldn't keep up good relationship with his professor. They quarrelled and as a result Vesalius started working independently.

He had ideas of his own which were different from the conventional ideas of the time. Working independently he was able to put them into practice. His reputation rose so high that students came to study his techniques of dissection. But his quarrel with Sylvius and the rest of the Faculty of Medicine continued. Their differences were mostly about Galen's teachings with which Vesalius could not agree. Galen was held in great esteem by the medical profession in Paris as everywhere in Europe. His ideas were passed down through centuries and had become the unquestioned word in medicine and anatomy.

In 1536, war broke out between France and what was then known as the 'Holy Roman Empire'. Vesalius, as an alien became suspect and was forced to leave Paris. He returned to Louvain and resumed his medical studies there. Medical students were faced with a difficult situation at Louvain at that time. They were required to watch dissections of the human body but both the Church and the Government looked at it with disapproval and only a strictly limited number of corpses were made available to them.

But this didn't put off Vesalius. He stole corpses from graves and became an expert grave robber. Once he even stole a skeleton of a criminal that had been left rotting and swinging from the gallows and hid it under his bed.

He had to do a stint of military service, after which he went to Venice in Italy. There he joined the University as a lecturer. Again he angered the professors with his totally new methods of teaching.

In those days the practice of teaching anatomy was to engage a barber to do the dissection while the professor lectured from the platform. Vesalius dismissed the barber surgeon (barbers were also surgeons in ancient and medieval Europe) and taking the knife in hand, performed the dissection himself, at the same time lecturing to the assembled students in spite of the unpleasant stench. To lessen the stench dissections were usually held in the open air; even the stench was unbearable and the situation extremely trying for both students and lecturers. Because of this, winter was preferred for dissection classes, as bodies don't rot in the freezing cold.

After working for some time in Venice he moved to the famous University of Padua, also in Italy. There within two days of his admittance as a graduate student he was awarded his Doctorate in Medicine in 1537 , and was appointed to a full professorship. Thus Andreas Vesalius, at the age of of 23 became a member of the most prestigious medical faculty in all Europe.

It was at Padua that Vesalius demonstrated that many of the ideas propounded by Galen were wrong. He also showed that Galen's text on human anatomy, which had been the authority for European physicians during all those centuries, was not about the human body at all. It had been written on the basis of the anatomy of monkeys, but projected as that of man! Galen had perhaps worked on the supposition that the anatomy of a *monkey* didn't differ much from that of a man.

Vesalius display the the skeleton of an ape side by side with the skeleton of a man and pointed out 200 differences between them. In this way he conclusively disproved Galen.

In 1543 Vesalius published his famous work *De humani corporis fabrica*. It was the most accurate book on human anatomy up to that time. It had excellent illustrations of the human body done by a pupil of the great *Venetian* artist Titian. The printing was also meticulously executed; sparing no expenses, Vesalius had engaged the most reputed printer of those days.

But in areas other than anatomy, the *Fabrica* also, contains errors, as revealed by modern medical science. For instance the act of digestion is stated to be a type of 'cooking' of the food in the abdomen; respiration is for 'cooling the blood.' The fact was that Vesalius, outside the field of anatomy, was a student of Galen whom he respected despite pointing out the errors in his work.

Fabrica was his final contribution to medical science. Shortly after its publication he left the teaching profession and secured a position as physician to the Emperor Charles V, and later to Charles' son, the Spanish king Phillip II.

When returning from a pilgrimage, his ship was caught in a storm off the coast of Greece. The ship was badly damaged. Vesalius somehow managed to reach the shore of Zanti, an island, but he died there shortly after, in the year 1564.

Antonie Van Leeuwenhoek

(1632-1723)

The Beginning of Microscopy

Antonie van Leeuwenhoek was a pioneer in the field of microscopy. Leeuwenhoek, who used microscopes of his own design and manufacture, was the first to observe bacteria and protozoa.

No one knows when the first true microscope was invented. But lenses were used as early as the time of the Assyrian civilization which is much earlier to the Greek. Both Romans and Greeks were aware that lenses could be used to magnify minute objects. The Arabs used lenses. The 13th century English writer Roger Bacon wrote about the magnifying qualities of different lenses. In the 16th century, a Swiss naturalist Konrad Gesner, used lenses to study snail shells.

Probably the first of the compound microscopes in which a combination of lenses is used to get required magnification, was made by a Dutchman Zacharias Janssen around 1590. By the 17th century the use of microscope had advanced, and there were a few men who were using it to look into a world too minute for ordinary eyes. Leeuwenhoek (pronounced Lay-ven-hook) was the most remarkable of these 17th century 'microscopists.'

He was born at the picturesque old town of Delft in Holland in 1632. His father was a basket-maker and mother, the daughter of a brewer. A brewer is one who makes beer, and Delft was famous for its beer and its fine chinaware. His father died when he was still a boy and his mother remarried. Young Leeuwenhoek went to school at Delft till he was sixteen, after which he was sent to Amsterdam to become apprentice to a draper.

He completed his apprenticeship at the age of twenty-two and returned to Delft. There he opened his own drapery shop — a draper is one who sells cloths and curtains and similar items. He married in the same year, he had two children and his business prospered. He became a typical, successful, small-town businessman, a respectable citizen appointed to a variety of positions in the municipality.

> "My work which I've done for a long time, was not pursued in order to gain the praise I now enjoy, but chiefly from a craving after knowledge, which I notice resides in me more than in most other men."

Drapers used lenses to examine the clothing and other goods in their shops. It is not known when Leeuwenhoek shifted his lenses from those business items to other things. It must have been early, for the microscopes he developed were of superb craftsmanship and not of the usual type required by a draper.

His talents came to the notice of a noted physician and anatomist of the time, Regnier de Graaf, who wrote a letter to the Royal Society of London about him. In this letter, de Graaf said that there was an unknown draper named Antony van Leeuwenhoek, who built the best microscopes he had ever seen. Along with the letter de Graaf also forwarded a *letter* that Leeuwenhoek had sent him, describing some of his observations with the microscope. The observations were on mold, the mouths of bee, and the common louse. The president of the Royal Society was interested. He wrote to Leeuwenhoek asking for more details and sketches.

This was the beginning of a long correspondence between the amateur scientist and the Royal Society, the most important body of science in the world at the time. There were 372 letters in all, continued throughout Leeuwenhoek's long life - he lived till ninety-one - a 50-year correspondence which is a record in the history of science.

The letters were written in colloquial Dutch. They often began with some simple talk about life in Delft, or his personal habits, his pet dog, or his business upturns and downturns, and then moved on to describing an amazing variety of microscopic observations. He had little knowledge on how to present a scientific treatise in the formal manner, and often three or four different and unrelated studies were included in each letter. But the reports of his hundreds of observations were made with painstaking care and exactness and they made him the most famous and respected microscopist in the world.

Using his microscope, Leeuwenhoek made many important discoveries. Marcello Malpigi a microscopist like him and a contemporary had discovered the capillaries during his study of the wing membrane of bats. Leeuwenhoek's observations confirmed Malpigi. The discovery of spermatozoa in seminal fluid was another important contribution of Leeuwenhoek.

But above all, it was his discovery of living things so small that the naked eye can't detect them that astounded his contemporaries. Studying a drop of water under his lens, he was amazed to see that it was full of life. "They stop, they stand still," he wrote "and then turn themselves around with that swiftness as we see a top turn round, the circumference they make being no bigger than that of a fine grain of sand".

So the world was teeming with far more life than had been ever thought of! Living creatures existed everywhere.

> Leeuwenhoek was the first to observe the constituents of blood under a microscope.

Science was entering a new world of micro-beings, of bacteria, protozoa and so on.

In 1680 an honour, which every scientist covets, the membership of the Royal Society was given to him. He became famous. With fame a steady stream of visitors to his home also began, so much so that Leeuwenhoek regretted the loss of his privacy.

One day one of the visitors was Peter the Great, the Emperor of Russia! Leeuwenhoek had to carry his instruments and specimens to the canal as the emperor had come in a special yacht and didn't want to attract crowds.

He remained active till he passed away in 1723. His final letter to the Royal Society, mailed by his daughter after his death, bequeathed to the society, a cabinet containing 26 of his very finest and most beloved silver microscopes.

Archimedes of Syracuse

(287–212 B.C.)

The ancients, although their speculations and their theories have influenced the world in many ways through the centuries, did not, for the most part, bequeath to posterity many practical contributions to science. Archimedes, the Greek mathematician who lived in Sicily in the third century B.C., was a brilliant exception to this dictum. His work in geometry, hydrostatics and mechanics was of a pioneer nature, and the principles he laid down have been of paramount and lasting importance.

The heat of the Mediterranean day had gone. A well-dressed crowd thronged the busy main street of the Sicilian town. Men sat at the open cafes gossiping and sipping their wines or syrups. Sailors from the ships in the harbour swaggered along. Ox-carts rattled over the cobble-stones to the accompaniment of the drivers' shouts. A quiet gaiety pervaded the place, for war had not yet come, and peace and prosperity reigned. It was too early in the evening for any excitement or untoward happening such as might take place when the sailors' roll would become more pronounced, and the pickets would land from the ships.

But suddenly the smiling calm was disturbed. The carters' shouts were stilled. The rattle of the carts was stopped. The drinkers at the cafes placed their beakers on the tables, and rose to their feet. They gazed up the street towards the public baths where the sailors' guffaws and the ribald shouts of the commoner folk betokened something out of the ordinary. Questions leapt from mouth to mouth. Soon no explanation was needed, for to the unconcealed joy of the idlers and the horror of the few ladies who were taking the air, a man was seen running down the middle of the road, blindly, taking no account of obstacles—stark naked ! He was shouting something—one word—one word which he repeated over and over again.

Foreheads were tapped. A madman, obviously, suffering from the effects of the day's sun. The man was recognized by somebody among the well-dressed idlers. His name passed from lip to lip. It was a well-known name—the name of the greatest mathematician and mechanical engineer of his time, the greatest then known to the civilized world.

Long after he had passed busy tongues speculated on the happening, but the true explanation of the great man's eccentric behaviour did not leak out till the following day. Then the story of the greatest feat of absent-mindedness known to history was told. It has been told many times since throughout the twenty-two centuries which separate us from the event, but it remains always worth telling.

The naked runner was Archimedes, Greek mathematician and pioneer in the domain of mechanics, who was born in the Sicilian town of Syracuse about two hundred and eighty-seven years before the birth of Christ. Here is the tale of why a venerable philosopher ran in the roadway without his clothes.

Hiero, King of Syracuse, had commissioned from a goldsmith of the town a crown of pure gold, but, having taken delivery of the finished article, he was suspicious. There was reason to believe that the craftsman had mixed with the gold a certain amount of other metal of inferior value. But how to find out ? There was no direct evidence, and it was therefore obviously a case for the learned men of the city. And who more learned than Archimedes?

The mathematician was therefore charged with the task which would nowadays be considered a simple one, but was then a matter for serious thought. Nothing known to science could be brought forward to prove fraud or otherwise on the part of the goldsmith.

It is more than probable that the human side of the problem interested Archimedes not at all, but the scientific puzzle worried him intensely. This worry pursued him everywhere he went for days, and persisted through the routine acts of his daily round.

In the normal course of that routine, he went to the public baths. We can imagine him standing at the edge of the bath tub as he prepares to enter it, absently allowing the water to flow until he cannot help noticing it. We may watch him lowering himself into the tub, then rising from it, studying with an interest and preoccupation, which seems childish, the changing level of the liquid, and all the while keeping his gold problem at the back of his mind, ever ready to push its way to the front.

Suddenly he splashed out of his tub, shouting at the top of his voice: "*Eureka! Eureka!*" (I have found it ! I have found it !). Without waiting, or even thinking of such a detail as clothes, he promptly took to his heels. Finally, Marcellus realized with natural chagrin, mixed with no little admiration for his enemy, that his engineers could not cope with the Syracusan, and decided to turn the siege into a blockade.

"Why fight with this mathematical Briareus?" he said, "who sits on the beach and baffles all our naval assaults as if they were a huge joke!"

Plutarch says of Archimedes at this period of his life: "In truth all the rest of the Syracusans were no more than the body in the batteries of Archimedes, while he himself was the soul which animated them. All other weapons lay idle and useless; his were the only offensive and defensive arms of the city."

A curious sidelight is thrown on the mentality of the inventor of all those deadly weapons by the fact that although his reputation had spread far and wide through his organization of the defence of Syracuse, he actually despised the ingenious mechanical contrivances which had made him famous. He carried this contempt so far that with one exception he refused to commit any record of them to writing. The solitary exception was far from being a warlike weapon, for it was a sphere which he made to imitate the motions of the sun, the moon, and the five planets in the heavens. Even the record of this is now lost, but Cicero more than a century later saw the apparatus and describes it.

The refusal of Archimedes to attach any value to his mechanical inventions does not seem to be due to any antimilitaristic or pacific tendencies, but rather to the fact that he regarded them as beneath the dignity of pure science. Perhaps Plutarch's explanation is another way of expressing the same thing: "He considered all attention to *mechanics* and every art that ministers to common uses as mean and sordid, and placed his whole delight in those intellectual speculations which, without any relation to the necessities of life have an intrinsic excellence arising from truth and demonstration alone."

That his work on these inventions was no more than a sideline dictated by the public needs of the moment, is evident from the list of those learned works from his hand of which existence is known to us. Who is the modern wrangler who would not be proud to put his name to a tithe of the following highly technical and abstruse essays, to which modern science owes so much:

1. *On the Sphere and Cylinder.* (A work in two books dealing with the dimensions of spheres, cones, solid rhombi, and cylinders, all demonstrated in the purely geometrical method.)
2. *The Measurement of the Circle.* (One book of three propositions.)
3. *On Conoids and Spheroids.* (One book containing thirty-two propositions.)
4. *On Spirals.* (One book containing twenty-eight propositions.)
5. *On the Equilibrium of Planes or Centres of Gravity of Planes.* (Two books.) This is the foundation of theoretical mechanics,

Aristotle's previous contributions being relatively vague and hardly scientific. In the first book there are fifteen propositions with seven postulates. The second book has ten propositions. Demonstrations are given, *practically identical with those employed at the present day,* of the centres of gravity of (1) any two weights; (2) any parallelogram; (3) any triangle; (4) any trapezium.

6. *The Quadrature of the Parabola:* This is a book of twenty-four propositions containing two demonstrations that the area of any segment of a parabola (i.e. a curve or conic section formed by cutting a cone with a plane parallel to its slope) is four-fifths of the area of the triangle which has the same base and equal height as the segment.

7. *On Floating Bodies.* A treatise in two books.

8. *The Psammites.* (Sand Reckoner.) A work dealing with the nomenclature of numbers.

9. *The Method.* An important treatise, discovered in 1906, in which Archimedes explains the mechanical methods whereby he arrived at many of his conclusions.

10. A collection of *Lemmas* or propositions in plane geometry.

This was probably not written by Archimedes in the form in which it survives, for it is a Latin version of an Arabic manuscript.

In addition to these, we know from quotations from contemporaries and later writers that Archimedes wrote a number of other books, now lost. The subjects of these were, naturally, mainly geometrical, but some treated of such matters as balances and levers, one dealt with astronomy, while yet another was an optical work, for a later author has quoted from it a remark on refraction.

It is difficult to estimate justly and accurately the merits of Archimedes as a math matician without a more complete knowledge than is available to us of the state of science as he found it. Certain it is, however, that he made discoveries on which modern mathematicians have based their methods of measuring solids and curved surfaces. He is, in fact, the only one of the ancients who has contributed anything of real value to the theory of mechanics and hydrostatics. This long-dead thinker did, indeed, play no small part in the making of our present-day mechanical civilization.

Like many other great men, his reputation with the populace was based on work by which he himself set no great store, but the cultured men of his own time and of later ages have been unanimous in considering him a pioneer and a genius. Their admiration for him was increased rather than diminished by the fact that by his works he made difficult things so easy that men wondered why they had not thought of those things themselves. "In fact," says Plutarch, "it is almost impossible for a man of himself to find out the demonstration of his propositions, but as soon as he has learned it from him [Archimedes] he will think that he could have done it without assistance—such a ready and easy way does he lead us to what he wants to prove."

Of Archimedes the man little is known. That sometimes he was strange in his manner is beyond doubt, and it is equally certain that he was often so absorbed by his scientific problems that he neglected his person and had even to be taken to the baths by force. It is recorded that during his ablutions he drew geometrical design in the soapsuds on his body.

He was, as has been seen in the case of his discovery of the principle which bears his name, childishly elated by each new discovery, but he was at same time modest and unassuming. Before publishing his work to the world, he invariably submitted it to his friend, Conon of Samos, whom he had probably met as a student at Alexandria, and whom he admired greatly as a mathematician.

Archimedes, the inspired engineer, even kept the *invader* Roman Marcellus at bay for nearly three years. In the end, it was weapon of treachery that succeeded where engines had failed, and Syracuse fell in 212 B.C. Its fall involved the death of its chief defender, but yet it was not the brilliant inventor of arms of war who was killed on that day; it was the absent-minded dreamer who on another occasion, years before, had run naked through the streets of his native town in the joy of a new discovery.

We do not know what problem it was which so engrossed him on the fatal day, for even the accounts of his death do not agree in details. We do know, however, that the victorious Marcellus had for his great opponent an unbounded admiration, and that he wished to meet him—not to crow over him, but to express the esteem in which he held him.

It is not difficult to imagine the final scenes. Marcellus, in midst of his other preoccupations, must have wondered why the great Archimedes was not among those who submitted to him. It could not have occurred to him that the man who had immobilized the Roman army and navy for three years was even then seated on the floor of his house, entirely forgetful of events outside, even unaware that the city had been captured. We can visualise that Roman as he gives an order to a member of his staff :

"See if you can find Archimedes for me! I should like to see him !"

We can hear the repetition of that order as it is passed on to subordinates in true military fashion :

"Find Archimedes! The General wants him!"

A soldier finds him. He sees an old gentleman seated on the floor, drawing funny lines on the tiles.

"Come on, you!" he orders. "The general wants you!"

The elderly gentlemen pays no attention—has not even heard. The order is repeated.

"Eh? What? Oh, go away. I'm busy!" That is all the soldier gets for an answer.

And so he takes a pace forward. He has his orders, and he will execute them. Archimedes is to be taken to the general. Is a silly game of noughts and crosses to interfere with military duty? Not so! He catches hold of the old man's arm. Archimedes shakes him off, mildly but firmly.

"I really cannot go until I have finished my problem."

This is insolence. There is only one way to deal with it. Angry flush mounts to the Roman's cheek. A last warning. It is unheeded. A thrust of the short, broad sword, and Archimedes is on top of his unfinished problem—dead through absent mindedness. And the life-blood that fed one of the world's most marvellous brains poured itself away.

Marcellus, whose orders had been thus tragically interpreted, was grieved to the heart at the death of Archimedes. " He turned his face away from the murderer as from an impious and execrable person," and seeking out the great man's relations, he bestowed special favours on them. The sage was given an honourable burial, and at his own request, expressed long before, his tomb was marked by the figure of a sphere inscribed in a cylinder, as he regarded his discovery of the relation between the volumes of a sphere and its circumscribing cylinder as his most valuable achievement.

After the death of Archimedes, his reputation was kept alive in the memory of men of science and letters, but his native city of Syracuse fell upon evil days, and it was more than a hundred and thirty years later – in 75 B.C. – that Cicero, then Quaestor in Sicily, found the tomb of Archimedes near the Agrigentine gate, overgrown with briars and thorns.

"Thus", the great Roman orator says sadly, "would this most famous and once most learned city of Greece have remained a stranger to the tomb of one of its most ingenious citizens, had it not been discovered by a man of Arpinum."

Birbal Sahni

(1891–1949)

The Sahnis belonged originally to Dera Ismail Khan, an important trading centre on the banks of the Indus river. Due to reverses in the family business they moved to Bhera, a small town in West Punjab, now in Pakistan. It was in Bhera that Birbal, the third child of his parents, was born.

Birbal's father, Ruchi Ram had had a difficult childhood. Yet, adversity and the loss of his father only made him more determined. Once, during a long journey, he had to spend the night on a tree. He did not have the money to go to an inn and he was afraid to stay anywhere else, in case somebody stole the bundle of books he was carrying. Later he became a professor at the Lahore Government College and made a name for himself. He also had the unique opportunity of working with Lord Rutherford at Cambridge. To a great extent, Birbal Sahni's scientific aptitude and his ability to face odds were inherited from his father.

In the early years of the twentieth century, the study of science was not considered a paying profession. No father wanted his son to take up scientific research as a career. Even an enlightened man like Ruchi Ram Sahni, a professor at the Government College of Lahore, wanted his bright young son, Birbal, to sit for the civil service examination. But Birbal had other plans. He was fascinated by plants and wanted to take up botany as his field of specialization. Not wanting to annoy his father, he had a frank talk with him. Ruchi Ram was easily persuaded. After all, he had himself sowed the love of nature in the boy.

The Sahni children used to go on treks to the nearby hills. Birbal's love for botany and geology can be traced to these trips, He grew up in an atmosphere of culture, education and nationalism. His sisters went to school and college along with him and his brothers. One of his sisters was the first woman graduate of Punjab University. Eminent political figures like Motilal Nehru, Chittaranjan Das and Gopal Krishna Gokhale visited their house at Lahore.

Birbal had his early education in Lahore. After school, he joined the government college there. Although Birbal's love of plants was partly inspired by the family herbarium, where plants used to be preserved in bottles, the man who gave him the necessary guidance was his professor at the Government College, Lahore—Shiv Ram Kashyap.

Like his siblings, young Birbal would frequently bring home a sapling which had caught his attention. Once he planted the Indian laburnum, also known as amaltas or golden shower. It was with great pleasure and excitement that the family watched it grow into a tree bearing beautiful yellow flowers. All his life, Sahni had a particular affection for this tree. When he built his house in Lucknow, rows of amaltas were planted along the approach road.

In 1911, Birbal went to Cambridge. He got a first class Tripos in natural science and won a scholarship. He did research under the well-known botanist Professor A.C. Seward, and was awarded a D.Sc. for his work on fossil plants. He was so well-versed in the subject of Indian plants that he was asked to revise a standard textbook of botany while merely a student. That book is still widely read by botany students.

He returned home in 1919 and joined the Benares Hindu University. He then moved to Lahore and finally to Lucknow where he taught and worked as professor of botany for the rest of his life. Lucknow was soon to become the most active centre of botany in the country. Sahni also took charge of the department of geology, because he felt that the two subjects were interlinked. He was made a Fellow of the Royal Society in 1936—the fifth Indian to get that distinction and the first Indian botanist. He often had to travel to various parts of the world—to museums, in particular—searching for specimens. Then these specimens had to be placed in the proper genera and species, a process similar to finding the exact solution to a jigsaw puzzle.

Paleobotany is the study of fossil plants preserved in rocks dating back to millions of years ago. Through the ages, plants have evolved from simple to more complex forms. First there were water plants; then land plants appeared during the paleozoic era. Records of the history of the world are contained in fossils. But since the fossil remains appear locked in rock layers, they are closely related to the geologist's area of investigation. The fossil plants indicate the age of the rock, and also point to facts regarding climate, temperature and topography. If a certain variety of plant is found, then scientists can identify the type of climate and soil needed to nurture such a plant. Plant fossils also suggest possible lines of evolution. So, the study of fossil records and the determining of the geological age of the earth go side by side. In the words of Sahni:

Imagine for example that in a coal mine on a certain day a person had been sitting by the trough eating grapes and throwing the seeds into the water, then we could easily tell the day of the week by the grape seeds in the particular layer of chalk that was then being laid down. Or if during a certain night there were swarms of insects around a light in the mine, and some of the insects, falling into the trough or washed into it by the streams, were buried in the chalky silt then being formed, we could exactly date and even time the layer by its contained insect remains.

Most of the known fossils of plants have been found in peninsular India. During the Mesozoic era, India was part of a vast landmass, of which present South America, Africa and Australia were also parts. Geologists have called this ancient continent Gondwanaland. An ocean separated it from the northern landmass. During the Tertiary era, Gondwanaland broke up, and the present continents were formed.

During the latter part of the Carboniferous age, a large part of the southern landmass become covered with ice and most of the greenery was destroyed. This has been deduced from the fact that glacial deposits found in India, Australia, Malaysia, South Africa and South America are somewhat uniform in nature. Later, the climate must have become warm. The climate and the vegetation during these different geological periods can once again be determined by the study of fossils. Sahni developed an interest in the study of Gondwana fossil plants from his Cambridge days. This is how it came about. Some specimens had been sent to his professor in Cambridge from India. The investigations were carried out jointly by both of them and the findings were published in a joint paper on Indian Gondwana plants.

Sahni's next important publication was on conifers. From a comparative study of fossil flora, he was able to show that there were important differences between the fossil flora of the north and the south of India and also between conifers of Europe and India.

Sahni made a list of all available fossil plants of the Gondwana continent. In this way, he provided valuable paleobotanic evidence to the theory of the continental drift. If one studies the eastern coast of South America and the western coast of Africa, a curious fact comes to light. The coastlines seem to match each other as if they had belonged to one landmass, had been wrenched apart, and then had drifted away. In fact, this and other evidence like similar plants and animals led Sahni and other scientists to the continental drift theory.

Sahni also did important work in proving that a particular type of plant, generally regarded as a fern, actually belonged to the class of seed-bearing plants. His many other researches helped him solve a number of geological mysteries. It was typical of Sahni to solve the problems of one branch of science by the methods of another. Specialization, he believed, tended to confine knowledge to compartments. One is apt to overlook the links between them and the fact that the developments in one can have bearing on another branch of science.

Sahni's method was unique. He attacked the theory of the continental drift in terms of fossil plants, that is, from a botanist's angle. He carried out researches on crop production during the Indus Valley Civilization and bridged the gap between botany and archaeology. In Harappa, he found charcoal remains of a kind of conifer. From this, he concluded that Harappa had trade links with the people of the northern mountain region, where the conifer grew.

In Khokrakot, which is near Rohtak in Haryana, he discovered shapes of rice husks impressed on clay. He recovered cells and stomata from the clay. The evidence was strong to suggest the existence of this particular type of rice here two thousand years ago. Even the name of the city 'Rohtak' could have been derived from *rohitaka*, a short evergreen tree possessing medicinal qualities. The tree does not grow in Punjab or Haryana any more, though it is still fairly wide-spread in north-eastern India, the Western Ghats and Malaysia. Sahni even went to the length of establishing that man had appeared in India before the Himalayan range existed.

All this points to Birbal Sahni's wide range of interests. For him botany, archaeology and geology became one. His contribution to Indian numismatics, that is, the study of coins, is also considerable. His discovery of an ancient coin mould in Khokrakot was followed by a study of the techniques of casting coins in ancient India'. He then collected data from other countries and made a comparative study of methods of casting. He came to the conclusion that at least a century before the Roman era, India had developed a fairly sophisticated type of coin mould. The results were published in 1945, in an article called 'Techniques of Casting Coins in Ancient India'. Noted indologists were thrilled by Sahni's discovery. They had had no idea that mints had existed in ancient India. Encouraged by the results of his own investigations, Sahni examined the seals found in Surat and Ludhiana, and came to the conclusion that they were not seals, but coin moulds.

Sahni was deeply involved in the progress of scientific knowledge. In 1924, he became the founder-member of the Indian Botanical Society and was elected its president more than once. In 1935, he was made the vice-president of the paleobotany section at the International Botanical Congress in Amsterdam. He was president of the National Academy of Sciences twice.

After Independence, he was offered the post of secretary to the Ministry of Education. Since it was a request from Maulana Abul Kalam Azad, Sahni found it very difficult to refuse. But the idea of leaving his laboratory upset him so much that he sent a telegram to the Education Minister explaining why it was impossible for him to accept the offer.

Through his work, Sahni came to be known as one of the foremost paleobotanists in the world and certainly the subject's pioneer in India. He led a very active life till the last. A few weeks before his death, he made a trip to the Raj Mahal Hills in West Bengal to collect material for research. It was through Sahni's efforts that the Institute of Paleobotany at Lucknow was established, though he did not live to see the building completed. He died on 10 April 1949, a week after the foundation stone was laid. The foundation stone has embedded in it seventy-seven fossil specimens collected from all over the world.

Charles Babbage

(1791–1871)

The Father of Computers

With his eyes closed and mathematical tables open before him, young Charles Babbage was lost in thought. When his friend entered the room and saw him, he asked, Well, Babbage, what are you dreaming about?" Babbage replied, "I am thinking that all these tables might be calculated by machinery". That was how the idea of computer occurred for the first time to Babbage. However, it was soon forgotten and occurred again to him several years later when he and his friend had become tired of laborious calculations. He then wished if he could build a calculating machine that ran on steam power, a novel, emerging driving force in those days. And he went on to turn his idea into a working machine from the scratch. But, struggled as he throughout his life, he never became successful. Nevertheless he is today considered to be the undisputed inventor of computer because he conceived of it in totality, a feat never achieved before him.

Babbage was born on December 26, 1791, in Surrey, England, to a wealthy merchant and banker. From his childhood he was curious to find out facts for himself. Once he entered an abandoned garret in the evening to see ghosts and devils. At the age of 16, he thought of a device to walk on sea. In the process of testing it, he almost drowned himself. Meanwhile, the bug of mathematics had bitten him. He used to get up at 3 O' clock in the morning to solve mathematical problems. He joined the University at Cambridge but found the atmosphere dull and so gave up formal education. It was during his stay at Cambridge that he gathered like-minded young men interested in mathematics and formed the Analytical Society. It was the first society among several that he founded during his life-time, thanks to his friendly and gregarious nature.

His father always wanted that Babbage should take up a job because he had got married in the meantime and had children. But he never did so throughout his life and lived on his inherited father's property. He travelled almost the entire Europe, watching keenly the scientific scene in France and Italy and comparing it with that in his own country. He regularly threw parties at his home, which were attended by prominent scientists of the time, and also attended several hosted by others. He was curious to know not only about science and technology but also about commerce, economics and social science. He was an expert cryptologist. He invented a signalling system for lighthouses. Once he descended a bubbling volcano to understand its mechanism. He even wrote a research paper on how to pick a lock. He also developed a technique for lighting in theatre. A populariser of science through lectures, he always advocated systematic application of science and technology to industry and commerce. In fact, he went on to write a book on this issue which eminent economists like Karl Marx and John Stuart Mill quoted in their now famous works. He was a fearless crusader of science, too. In another book he criticized the way science was then handled in England and openly denounced some eminent English scientists for their corrupt activities. Once he even stood for a seat in Parliament. But while he was engaged in diverse activities his mind was always set to build a calculating machine which could save humanity from drudgery.

The idea of how a machine could perform calculations appeared to Babbage during his visit to Paris in 1819. Here he came across the impressive seven huge volumes of mathematical tables that had then been compiled under the supervision of a civil engineer, Gaspard Francois de Prony. In those days, such tables were often made by eminent mathematicians who used to labour for years. But de Prony's tables had been made in three stages by three different sets of persons. In the first stage, a small group of eminent mathematicians selected the necessary formulas for calculations. In the second stage, a slightly bigger group of mathematicians converted the formulas into numbers. In the last stage, about 60 persons, who knew nothing about mathematics except how to perform additions, subtractions, etc, correctly performed the laborious task of actually making the tables.

Babbage therefore thought that he could build a machine which would conduct different types of mathematical operations in its different parts and at different stages. He worked out the entire details of such a machine and became so excited by his success that he fell ill and was advised to take rest. He eventually built a working model of the machine which he called "Difference Engine". It was called so because it worked on a mathematical technique which repeatedly added differences between numbers to perform various types of calculations. Composed of wheels, axles, levers, etc., like the previous calculators, it was nonetheless more complex than any devised till then. Moreover, it did not need the assistance of a human being, off and on, and so was more or less free of human errors. It also rang a bell when a given task was over, and printed the results.

The working model of the Difference Engine became a sensation in the scientific world. So much so that it attracted the attention of the British Government. Seeing its importance in making accurate calculations for navigation, a must for a seafaring nation like England, the Government immediately gave Babbage a grant of $ 1,000 to build a full scale Difference Engine. However, for reasons still unclear, only a part of the Difference Engine was eventually built. Some historians think that the

Difference Engine could not be completed because the necessary technology was not available. In fact, Babbage had even to design and build the requisite tools for making a part of the engine. Others think that Government did not provide Babbage enough money to complete the engine. Besides, workers cheated Babbage. The Difference Engine remained incomplete, which can still be seen in the Science Museum, London. Nevertheless, it was a piece of engineering marvel of those times. If not anything else, it certainly produced a number of engineering tools which gave a boost to the then growing industries in England.

It was at a critical juncture in 1834 when the British Government had denied Babbage any more money for the engine and when he had lost his wife and two children, one after another, that the vision of a computer appeared to him. Having suffered so much due to the Difference Engine, any other person would have liked to wash his hands off any calculating machine. But his genius lay in the fact that he continued to pursue his dream. He called this machine "Analytical Engine" and began to work on the details. In fact, it is said that till his death at the age of 80, the Analytical Engine continued to occupy his mind. He was always adding new things to improve its working from purely theoretical point of view because the had no money to build one by himself.

The Analytical Engine has more or less the same working principle as a modern computer. Like a modern computer, Babbage introduced an Input device in the Engine. The Input device was a collection of punched cards. The way a card was punched indicated a particular number and the sequence in which the cards were kept indicated its program; the type of calculations that have to be performed on the numbers. He had adopted this idea of punched cards from the punched cards then used to weave a number of designs in the newly invented weaving loom. He visualised that the punched cards would perform different types of calculations, just as they weave a number of designs on a cloth in a loom. Seeing the way cloth was manufactured in a factory, he forwarded the idea of a "Mill" where numbers would be kept waiting, just as a cloth is kept before it is sent to market. He also forwarded the significant concept of feeding back into the engine the results of calculations for use in subsequent calculations, like "the engine eating its own tail". This concept was not much exploited until John von Neumann provided a sophisticated form of it in "Stored program concept" much later. Babbage also introduced "Control Unit" whose function would be to control the various operations of the engine. It was indeed a marvellous blueprint of a modern computer from a man when a steam engine was looked upon like a Space Shuttle today. Obviously, few persons could appreciate the significance of the engine and still less understood its working.

Among those few persons who understood the value and working of Analytical Engine was a charming young lady, Ada Lovelace, daughter of poet Lord Byron. She encouraged Babbage to build the engine and even induced him to bet money on horse-racing to collect money for it. He also tried writing novels to make money. But he met failure at all fronts. Nevertheless, he continued to forward applications for his engine in various walks of life, namely, industry, commerce, statistics etc., till his death in 1871. In his later years, he however came to know that two Swedish printers, George Scheutz and his son, had successfully built his Difference Engine after reading an account of it in a daily. Not at all jealous like any other inventor, he managed to secure them a Gold Medal at the Paris Exhibition in 1855, an honour he himself deserved.

Talking and Listening Computers

Anybody who has seen the wonderful science-fiction film *2001: A Space Odyssey* might remember the computer HAL which could listen to and talk of the astronauts in the spacecraft. In fact, such a type of computer has been a dream of computer scientists for a long time. It now seems near fulfilment. Although listening and speaking computers have already been installed at various airports in the United States, they have some serious drawbacks. They have a highly limited vocabulary and can misunderstand words if they are mispronounced or are spoken with a different accent. For instance, if the human being has a cold or cough, the computer would not be able to respond to the person's instructions which it does ordinarily. Or, instead of an American, if an Indian addresses some instructions, the computer would not be able to understand him. Experts, however, think that all this is likely to change soon. Computers would be able to respond to any voice spoken with any accent, whether the person has a cold or no cold.

When a word is spoken, it produces a unique pattern of waves — electrically speaking. In the memory of computer, such unique patterns of various words are stored, so that when any word is spoken it matches the patterns thus generated with the ones it has in its memory. It is, therefore, able to "read" the words spoken to it and follows the instructions accordingly.

What is the use of such a computer? It would easily be employed in various factories and offices where instructions have to be given from time to time to perform certain routine tasks. To date, instructions are given to a computer either by punching cards or typing words, which are comparatively cumbersome methods. Talking directly to the computer would be easy and fun. Such a computer would particularly be of great help to handicapped persons, who are unable to make use of their hands and perform any task.

C. V. Raman, Sir

(1888–1970)

Noble Laureate Physicist

Dr. Chandra Shekhara Venkata Raman was one of the most distinguished scientists of the 20th century. His discovery known as the 'Raman Effect' made a very distinctive contribution to the field of physics. For this discovery, he was awarded the Nobel Prize for physics in 1930. He was the first Indian scientist and also the first Asian to receive a Nobel Prize for physics. This discovery of Raman is significant for one more reason. C.V. Raman studied and conducted his scientific researches living in India while there was the widely shared notion that no great scientific discovery could be made under the then existing scientific facilities in India. The scientific talent of Raman appeared at a very young age. His first research paper, "The unsymmetrical diffraction bands due to a rectangular aperture" was published in London's Philosophical Magazine when he was only 18. This research made Raman famous in world's scientific circles. Later on, he made many important discoveries in light, sound and magnetism.

Dr. C.V. Raman was born on November 7, 1888 in Tirucharappalli (Tamil Nadu). His father, Chandra Shekhar Aiyer, taught physics and mathematics in a college so Raman had the advantage of a scientific background. His mother Parvati was a cultured lady. Raman was very intelligent since his early childhood. He passed his matriculation when he was only 11. In 1904, when he was only 16, he passed his B.A. from the Presidency College, Madras, and was the only student to get a first class. He did his M.A. in Physics from the same college and broke all previous records. Then he appeared in a competitive examination of the Finance Department (Accountant General) in Calcutta, but he was interested in scientific researches. Prof. Ashutosh Mukherjee, the then Vice Chancellor of Calcutta University offered him a teaching post. Raman left his highly paid government job to become a professor of science. In 1914, a science college was established in Calcutta and Raman was appointed its Principal. In 1921, he was awarded the prestigious degree of "Doctor of Science" by the Calcutta University and in 1924, he was elected a Fellow of the Royal Society of London. In 1928, Raman was elected President of the Indian Science Congress and in 1929, the British government in India conferred on him the title of 'Sir'. From 1933 to 1948, he was the Director of Indian Institute of Science, Bangalore. In 1943, the Raman Research Institute was set up by him in Bangalore using his Nobel Prize money. There he conducted research work for the rest of his life.

Meanwhile he continued to carry out different types of experiments and researches on the sun rays passing through water, transparent ice blocks and other media. For these experiments, Raman used a mercury arc and a spectrograph. Raman obtained some new lines in the spectrum on passing the sun rays through different substances. These lines were later called 'Raman Lines' and led to the discovery of 'Raman Effect'. The Royal Society of London awarded him 'Hume Medal' for the same discovery. He was also awarded Lenin Peace Prize in 1958. This discovery of Raman was, perhaps, the most valuable contribution in the industrial development of the world. The government of India also honoured him with the highest honour of the country 'Bharat Ratna' in 1954.

He died in 1970 in Bangalore leaving behind his wife and two sons.

Factoid
From Lab to Class—The Raman Effect !

How do you think C V Raman was earning his living while undertaking research that won him the Nobel Prize for physics in 1930? After his masters in 1907, Raman took the Financial Civil Service and topped the examination. He was posted as an assistant accountant general in Calcutta. He worked for 10 years in the Indian Finance Department, while continuing research in his spare time, in the laboratories of the Indian Association for the Cultivation of Science. In 1917, Raman was offered a Profesorship of Physics at Calcutta University, where he stayed for the next 15 years.

Charles Goodyear

(1800–1860)

The 'Rubber' Man

To invent rubber which does not become sticky during summer and hard like rock during winter was the mission of Charles Goodyear's life. In fact, he felt that God had sent him to earth to fulfil this mission! He was imprisoned several times because he could not repay the money he had taken for his experiments on rubber. Rarely does one come across a person with such a single-minded purpose in his life. For his obsession with rubber, he was even considered a harmless lunatic. At one time, when he was living in New York, U.S.A., a saying about him had become popular — "If you meet a man who wears an India rubber cap, stock, vest and shoes, has a money purse made of Indian rubber without a cent of money in it — that's Charles Goodyear"! However, the most tragic thing about this genius was that during his life-time nobody was ready to recognise his contributions to rubber industry.

Goodyear is today a famous tyre manufacturing company. In fact, it is one of the biggest rubber-manufacturing companies in the world. It has however nothing to do with Charles Goodyear! But it was named in recognition of his pioneering work on rubber forty years after his death. However, had Goodyear received immediate success with rubber, he would certainly have raised a company which would have become today as big as this one, because he belonged to a family of merchants and inventors. His father Amasa Goodyear was an inventor in his own right. He used to invent an item and manufacture it. For instance, he had invented a button-making machine, a new type of steel fork for agriculture needs. He was a prosperous merchant when Charles was born on December 29, 1800 at New Haven, U.S.A. Studious and hardworking, he used to occasionally attend to the family farms and hardware business. However, from childhood he was interested to become a priest but his father was keen that he should help him in his business. Eventually, his father's wishes prevailed and at the age of 17 he joined a hardware firm. Tall and handsome, he had a soft voice. As a salesman wherever he went, people liked him. But he was not interested in making money and used to yawn or think about something else when money matters were discussed before him. However, inventing new things thrilled him.

Goodyear married Clarissa. In 1826, he started his own business in Philadelphia. He opened a hardware shop which was the first of its kind in the U.S.A. Earlier, hardwares were always sold by wandering merchants carrying them on shoulders or horse-driven wagons. His business went on very well for some time and at one stage it looked as if he and his wife were likely to lead a happy and prosperous life. But three years later, a major illness forced him take to bed for some months and ruined him completely. He was too gentle to press customers who could not pay him. On the other hand, debtors began to hang about him. Goodyear thought he would pay off his debt if he made a new invention. He invented a few new things such as a new type of fork, a spoon-manufacturing process, a water tap, an air-pump, etc, but none brought him the success he desired. He therefore wanted to try something which would change his fortune. But debtors would not leave him in peace. Eventually, in 1830, when he was in the prime of his youth, the debtors managed to put him behind bars.

Goodyear had some good friends who managed his release from the jail. Thus began his efforts to make new inventions. He succeeded in improving an air valve. It was while searching customers for his valve in 1834 that he entered the shop of Roxbury India Rubber Company on seeing a life-preserver displayed in its show-window. The life-preserver, a rubber-lined jacked which could be inflated if the wearer is drowning, had a valve. Goodyear bought a life-preserver and incorporated his new valve in place of the old one. He went back to the shop and showed its salesman how the life-preserver became more efficient due to his valve. He wanted to bag orders for his valve. Unfortunately, he had entered the shop when it was suffering a huge loss. All the rubber products were melting in the summer heat. Waterproof rubber shoes worth millions of dollars had to be quietly buried because the rubber had melted exuding a noxious smell. The salesman told Goodyear that the entire rubber industry in the U.S.A. was facing 'a crisis at that time. If he could invent a new type of rubber rather than a new valve, the salesman told him, the company would be highly interested. Here was a challenging problem for Goodyear to solve and pay off his debts. But from this incident it should not be assumed that he blindly took up the cause of rubber industry and made it his life-long work. As a child, rubber had always fascinated him. He had even once dreamt of making clothes of rubber.

Rubber came to Europe through Spanish invaders who looted South America. The local Indians used to play a sacred game with a ball made of a sticky milk-material which oozed from a native tree. It was then called Caoutchouc (pronounced as "Koo-chook"). However, it was re-named rubber when the English chemist Joseph Priestley found that it could rub off pencil marks. The water-proof qualities of rubber were of special interest in Europe. In due course, chemicals such as terpentine, ether and naphtha were found to dissolve rubber. In England, a London coachman Thomas Hancock accidently discovered the process of "mastication" — of mixing rubber with various chemicals and producing a range of articles for common use. In the mild climate of the country, his products were not affected and so became very popular. But when the products were manufactured in the U.S.A. and sold, the summer heat melted rubber and the winter cold made it rock-hard. The fast rising rubber industry in the U.S.A. was suddenly caught in a crisis. It was at this stage that Goodyear had entered the shop of the rubber company.

Goodyear's laboratory was his wife's kitchen. His equipment consisted of pots and pans, a rolling pin, knife and raw rubber, apart from chemicals. As soon as his wife had completed preparing the food for the day, he would take over the kitchen for his experiments. Like a baker, he used to prepare a dough of rubber after mixing it with chemicals, knead it under a rolling pin to make sheets, and then cut it with a knife. Off and on, his wife and children used to become his assistants. In his home, one could literally see rubber all over — window panes, table-tops, dinner plates and vases, etc. In the beginning of his experiments, he unknowingly offended his neighbourers because rubber produced foul smells. But in course of time, he shifted his laboratory to some distance form his home. Every day he used to walk for miles together with a baggage of chemicals on his shoulders to reach his make-shift laboratory. Off and on, he felt he had achieved success and used to manufacture products of rubber but met failure. Having no other source of income, he had often to sell or pawn his household items for his experiments. Sometimes, friend also gave him money to support his family. And all the while, debtors were after him. He was put behind bars several times. But he was always happy because he could perform his experiments undisturbed in his prison cell.

It was during one of his jail terms that Goodyear made an important discovery. He found that if magnesia is used instead of terpentine to dissolve rubber, it produces a white rubber instead of a dark one. He called this "tanned rubber" and exhibited it after his release from jail at the New York American Institute and New York Mechanic's Institute fairs. He was not only awarded medals at these fairs but he also won the admiration of several businessmen. It was soon after this recognition that he began to wear an all rubber costume for which he became famous all over New York. However, the tanned rubber turned out to be tanned only skin deep. One day a rotten apple fell on the rubber spilling its raw acids and melting it. After trying all kinds of chemicals, even sugar, salt, sand, and so on, Goodyear thought that terpentine could be the culprit in that it must be responsible for melting rubber in summer. He therefore replaced turpentine by a solution of quicklime which is supposed to dry things. But this change had no effect on rubber, though it then became possible to produce rubber in sheets. It was the first time that rubber sheets could be produced in the U.S.A.

Once, Goodyear was painting a rubber curtain as was his habit of making his goods attractive to buyers. He made a mistake somewhere and to rub it off he used sulphuric acid. The acid, though rubbed off the paint, changed the colour of the rubber spoiling the beauty of the curtain. In anger he threw down the curtain. However, the next day, something struck him and he examined the spoiled curtain again. He found that the acid had made the rubber hard. Could the acid be the secret agent he had been searching for? Immediately, he began a series of experiments and found that acid indeed made the rubber more resistant to heat. But he could not then deduce that the secret lay in sulphur as he came to know later. During his series of experiments with sulphuric acid, he would have almost died because he used to close all the windows while performing them due to the fear that smell would offend the neighbourers. For about six weeks he almost struggled with death. Nevertheless, he eventually developed the process which he called "Acid-gas" process and patented it in 1837. He managed to convince a businessman William Ballard, whom he had met at the New York fairs, to provide him the facilities of his closed rubber factory so that he could manufacture rubber goods on a large scale. But as luck would have it, before Goodyear could show his results, the businessman went broke and gave up the entire idea.

In 1838, Goodyear went to live in the small town of Woburn, where there was closed rubber factory. Here he came across Nathaniel Hayward, an illiterate yet capable worker of the factory. Hayward told him that he was able to whiten the rubber by mixing it with turpentine and sulphur. He was however not sure whether it made rubber heat resistant or not. Goodyear felt sure that it would do so and decided to manufacture rubber goods. Even the U.S. Government placed a large order for mail bags to him. However, in course of time, this venture also turned out to be a flop as rubber melted in summer heat. Not only Goodyear again became penniless but he even lost his reputation in the rubber industry. His friends advised him not to work on rubber again and instead look after the well-being of his family. They assured him help if he wanted to start any other venture. But Goodyear was adamant.

His family returned to New Haven, while Goodyear decided to continue his experiments at Woburn. Every day, he used to walk about eight kilometres from New Haven to Woburn to perform his experiments and another eight kilometres back. It was during his worst days of life in 1839, when his family was on the verge of starvation, that he hit upon the secret which has made him world famous today. He had boiled rubber in terpentine and sulphur and was transferring a piece of it to another plate when it fell down into the red open fire place. When he took out the piece, he was amazed to find that it had not melted as he had expected it to be. It had instead charred like leather. He however saw a line of what he called "perfectly cured" rubber between the charred portion. He jumped with joy, showing his discovery to his brother and daughter, who were then visiting him. But all that he got were cold looks because his family and relatives had got fed up of his discoveries which came to nothing. However, Goodyear himself was convinced that he had at last discovered the secret. He had begun to feel for a long time that what he had not been able to achieve systematically would certainly be achieved by an accident with the grace of God ! He had never thought earlier of heating rubber because he knew it would melt at a very low temperature. But the accident revealed that rubber became more heat resistant without losing its flexibility. That very day he hung the rubber on a nail outside the door and got up early in the morning to see whether the night chill had affected it in any manner. The rubber, which he called "fire-proof gum", stood the test. Goodyear had discovered a process which was to revolutionise the entire rubber industry the worldover.

Although Goodyear was happy at making the discovery, he himself knew that he had to solve several problems before he could manufacture articles of rubber for mass consumption. The problems were mainly three: How much sulphur was to be mixed with rubber ? How much heat was to be applied to the mixture ? and how long to heat to produce excellent rubber ? The last two questions were difficult to answer because in those days the necessary heat controlling devices had not been invented. He went all over Woburn borrowing hearths on a part-time basis. He heated the mixture in open fires, applied steam heat, boiled it in sauce pans, roasted it in ashes, kept it in hot sand, so on. All techniques of slow fire and quick fire were employed. Through this series of hit and trial experiments for about three years he was eventually able to determine the amount of heat required to produce excellent rubber. Meanwhile, he himself had to undergo a series of trials. As he had sold out everything, even his personal books and his children's text-books to conduct his experiments, his family was on the verge of starvation. The severe winter in one year aggravated their situation. Fortunately, he came across a businessman, who had suffered heavy losses in rubber industry. When Goodyear showed him samples of his rubber, he was impressed and gave him a large sum of money to feed his family. One day, he went to the nearby city, Boston, with the hope of securing money for his venture but everybody turned down his offer. He therefore, could not pay the hotel bills and was put behind bars. After a few days, when he returned home, walking all the way in a storm, he found his son dying and he had no money to buy a coffin. Eventually in 1844, Goodyear started a new rubber-manufacturing company "Naugatuck India Rubber Company" with the help of his business friends. The venture succeeded. Goodyear made a variety of rubber articles, namely, tents, power and conveyor belts, printer's roller, pump valves, rubber bands, engine hose, pneumatic cushion, tyres for wheelbarrow, etc.

In due course, other rubber companies also copied Goodyear's techniques in the manufacture of rubber but none was able to produce as good rubber as his company could. One of his rubber products also reached Thomas Hancock, who after examining it deduced the way it had been manufactured. In fact, one of his associates, William Brockedon called the heating of the rubber-sulphur mixture "Vulcanization" after the Roman God of fire, Vulcan. The term became so popular that Goodyear had to give up calling his rubber 'fire-proof gum'. In the meanwhile, he had become a wealthy man but, unfortunately, he had imbibed such an obsession with the process of vulcanilzation that he could not tolerate any body else claiming priority for its discovery. Not only Thomas Hancock began to claim that he had discovered the process entirely on his own, even an American claimed priority and wanted to patent it. Goodyear spent a considerable amount of money to fight the case in the court, which went on for about eight years. Eventually, in 1852, the U.S. Circuit Court gave the decision in his favour.

In 1851, Goodyear participated in the Crystal Palace Exhibition held at London. His exhibit "Goodyear's Vulcanite Court" won the admiration of every one. It consisted of three rooms, whose walls, roof, furniture, carpets, draperies, etc, were made of rubber. There were also rubber combs, buttons, musical instruments, balloons, etc, on display. His exhibition surpassed Hancock's in all respects. Four years later he also displayed his exhibit in the Great Exhibition held at Paris. He had participated in the exhibition in anticipation of getting orders for his products but none came and he had no money to pay to the exhibition authorities. He was therefore put behind bars. Only when Napolean III awarded him the Grand Medal of Honour and the

Cross of the Legion of Honour that he was freed from the jail. Later, he also showed Napolean III his latest product *"pontoons"* — rubber boats which came into use only in the Second World War. One can imagine how far Goodyear was ahead of his time. In fact, he had also envisioned the use of rubber in making notes, flags, jewelry, sails for windmills and ships, and even rubber ships.

Goodyear however could not stand continous crises in his business and failing health. He died on July 1, 1860 one day after he heard the news of the death of his daughter. When he died, he had the satisfaction of having been given the priority of inventing vulcanized rubber, of having 60 patents on rubber to his credit, and of seeing an ailing rubber industry stand on its feet. However, his children considered him to be a big failure because he left behind a debt of $ 200,000. In fact, they had developed so much aversion for rubber that they were not ready to continue the rubber company he had raised.

Vulcanized rubber — or the first modern plastic? — has found so many uses in all walks of life today that it would not be too much to say that modern life is not possible without it. Goodyear's life-long struggles were indeed worth it.

Charles Parson

(1854–1931)

With their colourful flags on masts fluttering in the wind, an array of battleships lined the port at Spithead. England, in 1897. It was the naval review in honour of Queen Victoria's Diamond Jubilee. Suddenly, a small boat appeared and sailed fast through the ranks of the battleships. The astonished Commander-in-Chief of the Navy opened his eyes wide, and ordered a picket-boat to give a chase to the intruding boat. But the picket-boat could not catch up with the speed of the intruding boat let alone capture it! When the Commander-in-Chief examined the flag of the speeding boat through binoculars, he read "Turbinia". That reminded him of turbines. He wondered whether the boat was not being run by that crazy invention. He was not wrong. The 44 ton Turbinia was the first boat to run on a steam turbine. The inventor of the turbine was Charles A. Parson, who used it not only for driving ships but also for generating electricity. Within 20 years he brought the development of steam turbines to such a stage that even in these days of diesel engines and electric motors, they are used for many purposes.

Born on June 13, 1854, in London, Charles A. Parson was the son of an eminent astronomer, who was also a member of the British Parliament and President of the prestigious Royal Society, London. From an early age Charles had an inventive bent of mind, which blossomed in the environment that prevailed in his house. Besides meeting frequently the country's best scientists in his house, machines, furnaces and lathes were available to him. Once he and his five elder brothers built a steam engine to grind the surface of mirrors used in telescopes, which was earlier done by hands.

After doing his B.A. from Cambridge, Charles could easily have got the job of a Professor in a University. But he opted to be an engineer. He always dreamt of engines and wanted to build a new type of engine himself.

Parson joined as an apprentice to an engineering firm, where, apart from improving the steam engine, he wanted to invent rocket-driven torpedoes for sinking ships. But those were the days when the world did not need torpedoes but suitable steam engines to drive dynamos for generating electricity. The steam engines then in use were highly inefficient. Only 12 per cent of the heat of coal in form of steam was converted into mechanical work. This was because the linear to and fro movement of the steam engine's pistons had to be converted into the rotating motion of the rotor of a dynamo. Parson thought, why not build an engine which would directly produce circular motion needed for dynamo? He observed that some inventors had built steam turbines for producing circular motion, which were, however, highly inefficient and were not used.

A steam turbine works just like a water wheel or a windmill. A jet of high pressure steam is allowed to strike at an angle blades attached to the rim of a wheel, as a result of which the wheel is set into circular motion. Parson made a fundamental change in the design of turbines which made them two to three times more efficient than a steam engine. Instead of having one wheel in the turbine, he introduced a large number of wheels turning on one shaft. The jet of steam moved these wheels one after another till it was exhaused of its energy. In earlier designs, most of the energy contained in the steam was wasted. Of course, Parson did not arrive at final solution all at once. He first built small cardboard models of turbines, made them of iron and brass, and thereafter made small improvements over them to increase their efficiency — a method of inventing things he had acquired from his childhood. His first turbine-driven dynamo generated about 7.5 kilowatts (kW) of electricity.

Parson's turbine came to the notice of the public when an organiser of a skating show, after reading a small news item about it in a daily paper, asked him to install it for lighting the show. His turbine became so popular that an order was placed for a big turbine to generate electricity for a small town. He even received orders from Germany for his turbines. In the meanwhile, Parson started his own firm to manufacture turbines on a large scale to meet the rising demands. It was at this time that a friend reminded Parson of his age-old desire of building turbine-driven ships. After making some modifications in the turbine and the screw-propeller, Parson was able to launch *Turbinia* in 1897. It could cruise at the speed of 35 knots — the fastest in the world then. It was the most popular exhibit at the International Exhibition held in Paris in 1900.

The mischievous manner in which Parson brought the attention of the Commander-in-Chief of the British Navy to Turbinia did bear fruit. A few months later, the British Navy placed an order of building two turbine-driven cruisers. Unfortunately, both the cruisers met with accidents within a few weeks of their launch. The opponents of Parson put the entire blame on his turbines. For once, the future of turbines for propelling ships appeared bleak. A naval encounter in 1904 between the faster battleships of Russia and Japan, however, opened the eyes of British Navy Commanders. An order was immediately placed for building a big turbine-run battleship *H.M.S. Dreadnought,* which, when launched in 1906, created naval history. Subsequently, merchant and passenger liners followed suit. The 38,000 ton passenger liner *Mauretania*, which was built by Parson's firm won the "Blue Ribbon of the Altantic" and remained the fastest ship for no less than 20 years.

Few inventors have realised all their dreams during their lifetime as had Parson before he passed away quietly in 1931 while aboard a ship. During his life-time all the major power generating plants had replaced their piston engines by his steam turbines. The power generating capacity of his turbines had risen form the initial 7.5 kW to 200,000 kW!

The MHD Generator

A turbine generates electric power when it drives a dynamo due to the action of jets of hot steam under pressure. The turbine generator works in the following manner : coal or any fuel is first used to convert water into hot steam; the hot steam is allowed to move under pressure to push the blades of the turbine, which in turn, rotate the rotor of the dynamo to generate electric power. How about a generator which forgoes all these steps ? Such a generator has been invented, but to date it generates power on a small scale. It is called Magneto-Hydro-Dynamics or MHD generator. It has no moving parts like a turbine generator. It generates electric power when a special type of fuel is heated to a very high temperature under special conditions of electric and magnetic fields. It also generates electric power at a much cheaper rate as compared to turbine generator. In the days to come, it is therefore likely to replace the present turbine generators for the production of electric power.

Charles Robert Darwin

(1809–1882)

Interpreter of The Pattern of Life

I

The students were assembled in the operating theatre of Edinburgh University. The patient, a child, lay on the table. All was ready. The operation began. Suddenly one of the students, white-faced and on the verge of collapse, rushed from the room. No doubt there was every excuse, the year was 1826, twenty years before anaesthetics came to be used, when operations, however serious, had to be performed on fully conscious patients. But students, if they hoped to become doctors, had to face up to this. This particular student, however, did *not* hope to become a doctor. He had had enough already. At the only other operation he had attended he had behaved in the same way. Moreover, the ordinary routine, the bookwork and classes, bored him. He took no interest in any aspect of medicine. So, at the early age of seventeen, he had to abandon his medical studies.

His father, Dr. Robert Darwin, a huge and domineering man, six feet two in height, twenty-five stone in weight, was bitterly disappointed. He was a well-known doctor and the son of an even better-known doctor, Erasmus Darwin, who had been offered, but refused, the post of physician to King George III. It was naturally taken for granted that Charles should follow in their footsteps.

Charles had disappointed the doctor even before this. He had done badly at school and his reports had caused his father to remark, "You will be a disgrace to yourself and all your family." Charles's only interests in life seemed to be shooting and collecting. As the shooting season approached he could think of nothing else, and he was like a jackdaw for collecting. He collected anything and everything : shells, pebbles, stamps, coins, dead beetles, birds' eggs (though he had a solemn agreement with his sister never to take more than one egg from a nest), wild flowers, leaves, and so on.

His father regarded the collecting as childish waste of time, the passion for shooting he regarded more seriously. For of all types of men Dr. Darwin most disliked the idle, sporting type, and he saw his son developing into just that type.

That was why he had taken him from school and sent him at the unusually early age of sixteen, to begin his medical career.

Well, it seemed now to his father that the only opening left for Charles was the Church. Charles was not averse to the idea (it would have made no difference if he had been) ; as a country parson he would still be able to shoot and collect. Before becoming a parson he had to get a degree. He was sent to Cambridge to get one.

He thoroughly enjoyed his three years at Cambridge. He became a boon companion of the shooting, hunting, sporting crowd and finished up most days with jovial dinners, singing and card-playing. He attended only what lectures were necessary; but with the help of a tutor got his ordinary B.A. After that he had to spend another two terms at Cambridge.

Though he had allowed his sporting proclivities full rein, he was keen as ever on collecting. But he was narrowing it down giving it more thought and observation. His pebbles and bits of rock led him on to an interest in geology ; his shells, beetles, etc., to an interest in natural history. He sought advice and his keenness aroused the interest of his teachers, particularly that of Henslow, Professor of Botany, and Sedgwick, Professor of Geology. They showed him what to read and took him on country expeditions.

Now about this time a ten-gun brig, **H.M.S. Beagle,** was about to sail round the world to make a survey of the lesser-known coast-lines and waters, and her commander, Captain Vitzroy, had gained permission from the Admiralty to take a naturalist (unpaid) with him. He asked Professor Henslow if he could recommend one. The trip would take two years. Not many graduates could waste so much time and money. Henslow approached Charles and told him he need not worry about his disqualifications.

The idea filled Charles with such enthusiasm as he had never known—not even before the partridge shooting season. A trip round the world visiting the places he was reading about in his geology and natural history books! Then his spirits dropped. What would his father say? And his father would not only have to permit the trip but finance it !

His father said what might have been expected. Now that Charles was at last able to start on a career he wanted to go gallivanting off on a pleasure trip for two years! Certainly not ! In despair, more to get consolation than from any hope, Charles went to his uncle, Josiah Wedgwood. Surprisingly, 'Uncle Jos' thought the idea a good one and even promised to talk to the doctor about it. Dr. Darwin had always admired Josiah as a shrewd business man and, more important, he liked him. Josiah told the doctor that in his opinion this trip would have a stabilising effect on Charles just at the time he needed it. It would be hard work and there would be no chances of getting into mischief under Navy discipline.

He won. Dr. Darwin agreed to the trip and agreed to pay Charles's expenses.

So it came about that, on December 27, 1831, **H.M.S. Beagle** sailed from Plymouth carrying on board an excited fledgling naturalist, who was shortly to be very sea-sick and who was to return in five, not two years.

II

The Galapagos Islands lie on the Equator, 500 miles west of South America. Formed of blackish grey lava, pock-marked with craters and lined with fissures, these Pacific islands are no place for a holiday. Except in the centre of some of the larger ones, the chief vegetation consists of thorn-trees, cacti and weed-like flowers. Everywhere is a faint, unpleasant smell. It is the home of crabs, lizards and huge tortoises. The latter weigh up to 400 lbs. and probably attain an age of 300 to 400 years, but they are now on the verge of extinction. Naturally, if they can help it, ships never put in there, but on September 15, 1835, a brig might have been observed heading for Chatham Island, the most desolate of them all. It was the **Beagle,** looking the worse for wear and minus her lifeboat, which had been washed away in a gale.

The **Beagle** had spent all this time surveying up and down off the east coast of South America and then up and down off the west coast until, after nearly four years, she had turned her bows west and started on the long voyage home.

In those days ocean surveying and sounding took a great deal of time so the **Beagle's** naturalist was often able to spend long periods on shore, making journeys into the interior, writing voluminous notes and collecting specimens. Trust Charles to collect specimens! He collected so many that they would almost have sunk the **Beagle** several times over had he not been able fairly frequently to ship them home from various ports. The specimens varied from huge fossils and slabs of rock to mice, insects, spiders and water animals.

But Charles was not now just collecting, he was beginning to think as well, and the more he collected the more he thought. He noticed that the species of plants and animals changed as he went from one part of the continent to another, and yet that they often possessed remarkable similarities. He paid as much attention to geology as to natural history. He was surprised to find deep-sea shells on the highest mountains, hundreds of miles from the sea, showing that these mountains had once been fathoms deep under the ocean. This and other things forced him to conclude that the land all over the globe was continually, though infinitely slowly, rising and falling, at one time connecting continents and islands, making possible the spread of animals and plants, at others separating them by great tracts of water.

And here, just for a moment, we must leave Charles to make a note of the attitude of the scientists of those days towards life generally. Everyone knows what a species is. The bear, for instance, has only a few—the brown bear, the grizzly bear, the polar bear, and one or two others. Moths and butterflies have very many more, as those who collect them, and even those who do not, are aware. Spiders, too, have thousands of species, many very much alike. Now, scientists in those days believed that each species had been specially created in the beginning by the Creator and that these species never changed and never would change. Practically everybody else believed this too. And, of course, species never *do* seem to change.

Charles was as firm a believer in this as anyone else, but after his truly vast examination of innumerable species in South America he was a little puzzled as to why the Creator had made so many that were so similar and yet dissimilar. It seemed a waste of creative power. If, of course, over a long period, species changed as they spread from one territory to another, or even changed while they remained in one place, it would explain a lot. But species did not change. Back now to the **Beagle**. After anchorage, Charles—bronzed and wiry, viewed Chatham Island without enthusiasm but he was soon ashore watching the crabs and lizards and coming unexpectedly upon two of the giant tortoises going along a path they had worn by continual passage. They viewed him without fear and almost without interest. Exuberant at being able to stretch his legs again, and fit for any prank, Charles jumped on to the back of one of them and whacked the hind part of its shell. It rose obediently and waddled forward.

Its pace was not much more than a yard a minute but all the same Charles fell off. He tried several times during his stay in the islands without ever managing to stick on.

But his interest in these animals was not confined to having 'donkey rides' on them. He had seen from the first that they were a species known nowhere else. And yet they were strangely like some of the tortoises found on the American mainland. Later, he found it was the same with the birds and plants of the islands. Half of them were peculiar to the islands and yet had resemblances to certain finches and plants of America. He was still more intrigued when he found out that the tortoises on the various islands were different, very much alike but definitely different species. And it was the same with many of the birds and plants.

To Charles it seemed that there could only be only one explanation: at a remote period the ancestors of these animals and plants had got over from distant places to the Galapagos — and there were several ways in which they could have done so. Once there they had gradually changed and branched out into a number of different species. Which meant that species *did* change. And if that were so, it was even possible that all life had originated from one, or just a few very small beginnings.

That is what he was thinking when the **Beagle** at last set sail from the Galapagos and the giant turtles—but he kept his thoughts to himself.

New Zealand, Australia, the Indian Ocean with its coral islands, South Africa, then the long run home. Charles was no longer enjoying the trip. He was sea-sick again, sea-sick and very homesick. "I loathe, I abhor the sea and all ships which sail on it," he wrote in his diary.

On the afternoon of October 2, 1836, the **Beagle**, her sheets worn, her bottom foul, sailed into Falmouth Harbour and anchored. Her naturalist wasted no time. He took the night coach for Shropshire and never left Britain again for the rest of his life.

III

Not far from Bromley, Kent, in the typically English countryside that is now fast disappearing, lay the village of Down, a few hamlets clustered round an old stone church. Quarter of a mile away was a country residence, a large, brick three-storey house surrounded by spacious lawns, flower beds, flowering shrubs and long avenues of limes. Here, every day, a tall, thin man pottered about attending to plants or visiting his pigeons and rabbits. He had a bald dome and a brow that jutted out over his eyes like a cliff overhanging caves below. No one could possibly have connected this oldish-looking man with the young, pleasant-faced Charles Darwin who had sailed on the **Beagle** twenty years before, but it was he.

It had been intended, you will remember, that he should become a clergyman on his return, but he had abandoned that idea years ago. This does not mean that he had fallen into those ways of sporting idleness that his father had dreaded — the trip on the **Beagle** had cured him of that once and for all. He had left England a mere dabbler in science; he had returned an experienced naturalist and geologist. The specimens and the notes that he had sent home from time to time had attracted the attention of several scientists of high standing, and within a few years he occupied a distinguished place amongst them. Many of them were his personal friends.

At first he had lived in London, taking on much work in connection with various societies, but soon he became affected by a mysterious illness which forced him to get away from London and live a quiet life. So he had found Down House and gone there with his family. And he stayed there all his life.

No one ever found out the nature of this illness. He looked fit, took long walks, and lived to old age, but half an hour's conversation would exhaust him, and even the prospect of a journey, however short, would bring on headaches, nausea and stomach pains. It was obviously a 'worry' illness. There was nothing to worry him in his home life, though there was in his work. He was engaged in very controversial matters and he shrank from attack and adverse criticism. He was not a fighter like his friend, Professor Huxley.

He was working now on the masses of notes he had made and the specimens he had collected during the voyage. He was busy, too, making further collections and studies. The result of this vast labour was to appear in a book he was writing, called *The Origin of Species.*

All his work so far had confirmed his views about the changeability of species, but one thing had puzzled him: species changed, but what *controlled* these changes? Why did new species 'take over' from other species, rising to predominance and then themselves giving place to others? In the case of domestic animals man, of course, was the controlling force. Charles kept pigeons and he knew that his pouters, tumblers, fantails and the rest had all come originally from one pair of wild birds. By selecting from the types he wanted, aided by variations, man had brought about the different species. But there was no selection in nature.

The answer to the puzzle, when it came to him, seemed very simple. There *is* selection in nature—'the survival of the fittest'. Any species that comes to have even slight advantages over other strains must in the battle for existence, inevitably oust those strains, probably to be ousted itself later. He called this process *Natural Selection*.

Twenty years had gone by and Charles was still working on his book. His friends urged him to publish it lest someone step in before him. He took no notice. His subject was a never-ending one really, for it was life itself. Probably *The Origin of Species* never *would* have been published had not Charles received a letter one morning from a man named A. R. Wallace enclosing a paper he had written. In this paper were summed up all the conclusions about species at which Charles had arrived after so many years, of laborious work. His friends had been right. He had been forestalled.

Charles offered to withdraw his own book in favour of Wallace. But it was obvious that he had arrived at his theories long before Wallace, and in the end the matter was settled amicably.

But there must be no more delay. Charles worked as never before. *The Origin of Species,* was published in 1859. It had a mixed reception, mostly hostile. The fireworks were to come later.

IV

They came the following June at a scientific meeting in Oxford. Scenting a row, so many people assembled (over 700) that a large hall had to be obtained to contain them. Charles (luckily for him) was too ill to go. In his absence Huxley bore the brunt of the attack that developed.

The one thing Darwin had been hoping to avoid was mention of man. Man, of course, was included in his theory of evolution but there was little data to go on at that time. The so-called 'missing links' had not yet come to hand. Needless to say, man *was* mentioned.

Speaker after speaker attacked Darwin. Then rose Samuel Wilberforce, the urbane Bishop of Oxford (disrespectfully known to undergraduates as 'Soapy Sam'). For half an hour he ridiculed Darwin and Huxley, then turned to Huxley and asked him with an ironic smile if he claimed descent from an ape on his grandfather's or grandmother's side.

There was clapping and cheering. The meeting was almost solidly against Darwin. With dramatic timing Huxley delayed his reply until people were beginning to wonder if he was going to speak at all. Then he rose. He tore all the Bishop's arguments to shreds and finally stated (to give only the gist of what he said) that he would sooner have an ape for an ancestor than a man who, instead of attending to his proper duties, interfered in scientific matters of which he knew nothing: in other words that he would sooner have an ape for a foreparent than the Bishop of Oxford.

At this insult to a cleric pandemonium broke out. It looked almost as if the little group of Darwinites would be physically attacked. A woman fainted and had to be carried out. Amongst the shouting mob was Charles's old friend, Captain Fitzroy of the **Beagle,** waving a Bible aloft and referring to Charles as "that viper I once harboured on my ship". The Bishop sat with a strained smile.

As it happened, this disorderly meeting did good. It attracted publicity and brought the whole matter into the open. In ten years time there was hardly a scientist who did not believe in evolution. Most of the clergy, too, came to reconcile it with their religion.

V

In a kind of general knowledge test a class was asked, "What do you know about Darwin?" and a bright youngster replied, "He discovered evolution." More or less the same reply would be given by most people. But it is incorrect. Darwin's own grandfather believed in a sort of evolution and had written about it. A man called Chambers had also written about it, while a Frenchman, Lamarck, a famous zoologist born sixty-five years before Darwin, had held the same views and had carefully expounded them. So the idea of evolution can be said to have been 'in the air' in the

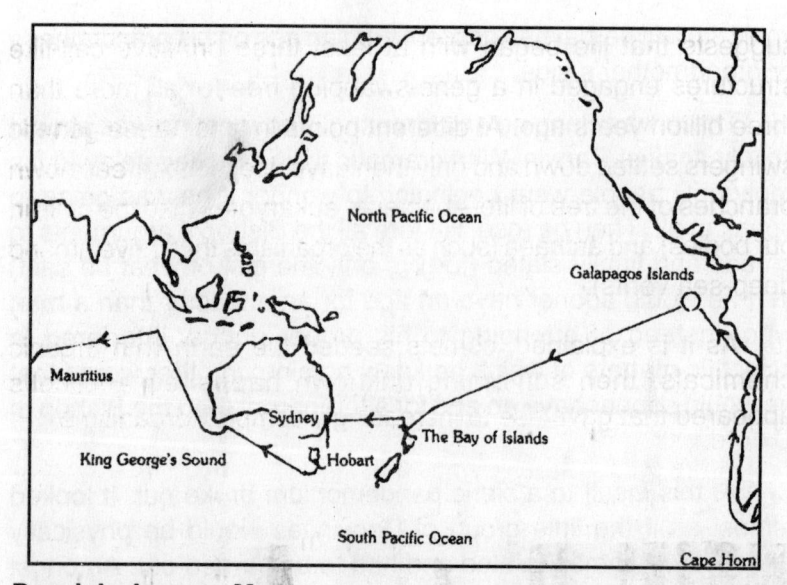

Darwin's Journey Map.
The sea-route followed by Darwin for circumnavigating the world aboard the ship **Beagle**. He completed this journey during 1831–1836.

19th century. Yet speculative ideas, unsupported by facts and upsetting to old beliefs, are easily ignored. Then Darwin came and by a mass of observations, a lifetime of careful research, proved that evolution was more than an idea—it was the pattern of life itself.

New Theory Fills in Gap Before Darwin

Forget what you learned in biology about the origin of life : that it began with a single mother of all cells and became increasingly complex was a simplistic notion anyway.

A New theory by evolutionary microbiologist Carl Woese, which may revolutionise notions on the origin of life,

Charles Darwin
Triple origin of the species

suggests that life began with at least three primitive cell-like structures engaged in a gene-swapping free-for-all more than three billion years ago. At different points in time, these genetic swingers settled down and only then gave rise to the three known branches of the tree of life—bacteria, eukaryotes (like the cells in our bodies) and archaea (such as the organisms that thrive around deep-sea vents).

As it is explained, comets seeded the earth with organic chemicals, then something unknown happened and cells appeared that gave rise to increasingly complex organisms.

Christiaan Barnard

(1922–2001)

Pioneer Surgeon

Few of those who knew Christiaan Barnard were surprised that the pioneering South African heart surgeon died at a European beach resort on September 1, 2001 while reading one of his own books.

The circumstances of his death illuminated the many ironies and paradoxes in the life of a poor Afrikaner farm boy who made his mark on the history of medical science with the world's first heart transplant.

Dr. Barnard (born on November 8, 1922) was the son of an impoverished farmer near the town of Beaufort West in the Karoo. He was personable and bright, earning a coveted place at the University of Cape Town, where he studied medicine.

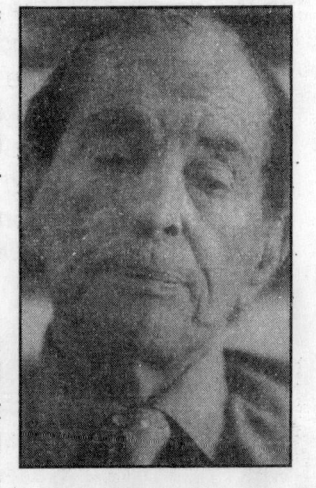

He advanced to specialised studies and became a resident at the city's Groote Schuur Hospital before winning a scholarship to Minnesota, USA to work under two distinguished heart surgeons. He returned to Cape Town where, with the help of a lung-heart machine donated by America, he developed one of the best heart surgery units in the world.

He had written papers about the possibility of heart transplants, pointing always to the difficulty of the timing in finding a donor. On a December night in 1967 he and a team of surgeons at Groote Schuur took the heart of Denise Darvell, 25, a motor accident victim, and stitched it into Louis Washkansky, 53, a grocer. Washkansky died 18 days later from pneumonia brought on by drugs designed to suppress rejection of his new heart. But already Barnard was being hailed as a pioneer.

He went on to perform several more heart transplants, more successful than the first. The heart unit at Groote Schuur became the epicentre of transplant surgery and Dr. Barnard's work, driven and relentless, won him the acclaim of every medical authority, as well as creating international acclaim for South Africa.

His research interests were wide-ranging, if seldom attentive to questions of ethics. In 1960, he transplanted a second head on to a dog. In 1974, he was the first to demonstrate a technique to give heart disease victims a "piggy back" second heart, and in June 1977 became the first surgeon to transplant a live animal heart — a baboon's — into a woman aged 25 who died shortly after. Her husband claimed that if the operation had not been carried out she would have lived. Barnard retorted that he performed the operation as a doctor — to save a life. But of heart transplants in general, he said: "For me, the goal of medicine is not the prolongation of life; it is improvement in the quality of life that is important." He was also an enthusiastic supporter of euthanasia.

On January 2, 1968, less than a month after his first success, Barnard performed his second heart transplant. The recipient was Philip Blaiberg, a retired dentist aged 58, and the donor Clive Haupt, a 24-year-old coloured man who had died from a stroke. The transplant of the heart of a coloured man to a White in South Africa aroused racial controversy. Blaiberg was spared steroids — which had weakened Washkansky — and kept in a sealed suite of rooms. He left hospital 74 days after the operation and lived for 19 months.

By then, surgeons were in dispute. Many felt that knowledge about overcoming the body's resistance to foreign organs was not advanced enough; others feared that some donors might not be dead at the time of transplants. Barnard dismissed these theories as ridiculous.

Barnard himself was not averse from the acclamation he received and responded with boyish enthusiasm to his star status. "Anyone who dislikes publicity must be mad," he said. He was received by the Pope, chatted with Diana, Princess of Wales, and featured several times in *Hello!* magazine. He was invited to tour the world, not only lecturing on heart surgery, but to talk and give interviews about subjects he knew nothing about. He obliged readily.

As he grew older, he became obsessed with the idea of prolonging youthful activities. He described ageing as "abnormal" and conducted research into injecting animal embryo cells into older people to restore their fading body functions. Barnard admitted to having several injections himself. The medical world, having acclaimed him, became increasingly alarmed as he gave his name to highly questionable balms and techniques of preserving youth, and became involved with a "rejuvenation therapy" clinic in Austria.

In 1986, Barnard sang the praises of an "elixir of youth" ingredient known as GSL or glycosphingolipids. Barnard argued that the product was sound because its basic ingredient had been found to delay the ageing process in fruit flies. But the Food and Drugs Administration in the United States demanded evidence of its effectiveness in people, and it was withdrawn.

Thereafter, Barnard divided his time between looking after his farm in South Africa and undertaking research work in Oklahoma. He made no great intellectual claims, maintaining that "stupid doctors become surgeons– all we have to do is cut things out, put things in and sew things up". His favourite book was *Gone with the Wind*.

He was awarded numerous prizes, fellowships and honorary degrees. His publications included *Surgery of Common Congenital Cardiac Malformations* (1968), *Heart Attack, You Don't Have To Die* (1971), *The Arthritis Handbook* (1984), four novels, and papers in scholarly journals. He died in Paphos, Cyprus; the cause of death was thought to be a heart attack.

Christopher Wren, Sir

(1632–1723)

An English Leonardo

I

In the year 1666 (one of the easiest dates to remember) the cry, 'London's burning!' echoed late one September night through the narrow, evil-smelling, timbered alleys of old London. We still echo that cry, "London's burning!" in an old song and go on to offer somewhat obvious but no doubt well-intended advice, 'Pour on water! Pour on water!' The anxious citizens, I am quite sure, did pour on water, hastening down to the darkly running Thames with what containers they could muster, but, the conflagration was far too fierce to be checked.

The fire, the Great Fire of London, continued for about a week. How it began we shall never know with any certainty for, in the face of such an overwhelming disaster, rumours spread as quickly as the fire. It was easy to say, and many people did, that the fire had been started by England's enemies (then the Dutch), anticipating the sabotage and war against civilians which, it is generally assumed, only our vile society has practised. Certainly atom bombers could hardly have done better. On September 8 some 200,000 refugees were encamped among their few remaining goods and chattels in the open fields 'towards Islington and Highgate'. Below them were the still smouldering ruins of London; 13,000 houses had been destroyed, together with the Guildhall, the Customs House, the Royal Exchange, the halls of 44 City companies, 87 parish churches and the cathedral church of St. Paul's itself.

One man, then thirty-four years of age, must have contemplated the destruction with much the same feelings of compassion mingled with a challenging sense of opportunity as ages ago, Noah, surveyed the old world after the havoc of the Flood. He was Christopher Wren, who at the early age of twenty-one had been appointed Surveyor-General. A new, magnificent London could now arise, phoenix-like, from these smouldering ruins and for this splendid city Wren already had his dreams. Instead of the huddle of houses, the dark alleys, the dens and slums that had, only that year, harboured the most terrible of plagues, he foresaw long, open vistas, radiating from a new, central cathedral church of St. Paul's and, at end of the vistas, beautiful parish churches with their elegant steeples gleaming in the morning sunshine.

II

London, and the area around St. Paul's, suffered a destruction no less horrible than that of the Great Fire in the year 1940. For many years afterwards it was possible to wander among scenes of desolation and see broken walls crowned with ragwort, ruins full of dumped junk, old cans and bedsteads, and haunted by thin, poor, hungry cats. Happily today Christopher Wren's masterpiece and greatest memorial, St. Paul's, still stands: the sight would gladden his heart but Wren would have been contemptuous of the rubble still around nearly twenty years later. For, within a *week* of the Great Fire, Wren had quite detailed, wide-sweeping plans for restoration ready to lay before the King— King Charles II.

In one respect the City of London was uncommonly fortunate. Charles II may have had few virtues but he had interests other than 'Sweet Nell of Old Drury' : legend says he was catching butterflies with Nell when the Dutch ships were sailing up the Medway. I doubt it; over the Great Fire, and the problems of rebuilding his capital city, Charles II showed imagination and a great sense of civic responsibility. The measures he advocated, both for checking the conflagration and for relieving the confusion and distress by which it was followed, were intelligent, far-sighted and extremely courageous. To a taste for architecture he added an interest in town planning.

Christopher Wren, in those September days of 1666, when the stink of charred wood was everywhere, was exceedingly fortunate. Fate had given him the greatest opportunity ever given to any man, and he served a monarch who was likely to be both approachable and receptive to his grandiose proposals.

Had King Charles occupied the position of an absolute monarch (which otherwise it is exceedingly fortunate that he did not) he and Wren might well have rebuilt a London of supreme and lasting splendour. But that was not to be.

Then, as today, the men of vision had their wings clipped by 'the obstinate Averseness of a great Part of the Citizens to alter'; the red-tapers; the little, puffed-up men of brief authority; the downright greedy; the anti-social 'Unwilling to give up their Properties tho' for a Time only, into the hands of public Trustees,' all these bone-heads made sure that a splendid opportunity was allowed to pass and a duller, less impressive, compromise resulted.

Certainly tremendous improvements to London were effected in detail. '*Si monumentum requires, circumspice*' - 'For his monument, just look around you'—was the epitaph inscribed upon the tomb of Sir Christopher Wren in his own St. Paul's. No other man has left so great an impression upon a nation's capital: from the Monument to the Great Fire, to the pedestal upon which the statue of King Charles I stands at the top of 'his own Whitehall' : from the great St. Paul's dominating the City to the (alas !) shells of City Churches, blitzed but now restored, there are many lasting memorials to this remarkable man, Sir Christopher Wren, their architect, a 'man of heaven-inspired projects'.

Besides the great and noble cathedral, Wren was responsible for 50 new parish churches, 36 City company halls and a Customs House. For all this his fee was only £300 !

III

What manner of man was this great architect ?

The material on the life of Sir Christopher Wren is meagre but, when a story about him has survived, it always gives a consistent and vivid picture of a man who was remarkably gifted in so many ways, an inquisitive man, a man who loved to experiment. Said a contemporary, "I must affirm that since the time of Archimedes there scarce ever met in one man in so great a perfection such a mechanical hand and so philosophical a mind."

After the Great Fire there was the minor but difficult problem of clearing away the ruins quickly and safely. The great shell of old St. Paul's, in particular, presented many difficulties. Wren had been interested in, and experimented with, gun-powder for some time. He had carried out experiments upon the unlikely value of explosives for cleaning sooty and smoky chimneys, experiments I would dearly like to have watched from a safe distance. To the consternation of the City Fathers he proposed to blow up the fire-charred walls by using nicely calculated charges of gunpowder and his first attempt was highly successful.

The walls rose a few feet in the air and then subsided into neat heaps. But one day, when Wren was not on the site, his second-in-command thought he would have a go. There was an almighty roar and the citizens for hundreds of yards around were bombarded with bits and boulders. Wren was much reviled and it was made very clear to him that the use of gun-powder must cease forthwith. And here we see the temper of Wren's ingenious mind. He was not daunted in the slightest. Having enjoyed a classical education at Westminster School under the great Dr. Busby, he remembered that the ancients managed well enough without gunpowder; they used great battering rams upon the walls of besieged cities. So, within a few days, the dangerous walls were again coming down in clouds of dust before the assaults of mighty battering rams.

IV

Like so many of England's great men, Wren was a parson's son and was born in the rectory at East Knoyle, Wiltshire, on October 31, in the year 1632. His mother, Mary, died when he was a little boy and his elder sister, Susan, was as much a mother to him as a young girl could be. Susan married when Christopher was eleven, marrying William Holder, mathematician, and a man who was to have a great influence on his little and, probably, desolate, brother-in-law.

Christopher was to live for ninety-one crowded years: few lives have been so fully lived for he was an 'early starter'. (I hate the word 'prodigy' ; it suggests some repulsive and priggish, old-before-his-time swot of a schoolboy.) He left school at fourteen and spent some time in the study of anatomy with Dr. Scarburgh, then going on to Wadham College, Oxford. He was Professor of Astronomy at Oxford before he was twenty-eight years old.

Wren was no dry-as-dust, absent-minded professor. He would have made the most superb young uncle for any lively schoolboy with an open mind and a thirsting curiosity about most things.

How wide were the man's interests !

Before he was twenty-one he was something of an authority on the anatomy of the brain. At the request of his king, Charles, he produced some remarkable drawings of insects microscopically enlarged, for he possessed great ability with his pencil. Then again he invented a 'planting instrument', the forerunner of the farmer's drill, a machine 'being drawn by a horse over ploughed and harrowed land that shall plant corn equally without waste'. At one time or another he engaged his active mind upon the problem of obtaining fresh water at sea, carried out experiments upon the purifying and fumigating of sick-rooms —at that time something well worth doing, and was among the first men to experiment in the transfusion of blood, succeeding in transferring blood from one animal to another.

Christopher Wren was the *English Leonardo da Vinci* : the complete man, the specialist in all things. Today our men of science follow the most narrow of lines: they do not even know everything about flowers, or insects, they know everything about *one* flower, *one* insect. For him knowledge was an ever widening funnel; today, unfortunately, the funnel points the other way towards narrowness.

In physics, medicine, anatomy, agriculture, astronomy, Christopher Wren was always experimenting, always asking questions and trying to find the answer. He worked for some time upon a weather-cock that would record, as it answered the winds, the story of the weather.

Wren was one of the moving spirits in the foundation of the Royal Society, a circle of inquiring men who met from time to time in his rooms when he was a teacher at Gresham College, London.

He was, indeed, first and foremost a man of science, his interest in architecture coming quite late in his life. Let a contemporary praise him: "As one of whom it were doubtful whether he was most to be commended for the divine felicity of his genius or for the sweet humanity of his disposition : formerly, as a boy prodigy, now as a man a miracle, nay, even something superhuman!"

Even allowing for the floridity of that florid age, that is praise indeed.

There is no question that Wren began the research or, in fact, made many vital inventions that others later exploited. Again hear the voice of his times, "I know very well that some of them" (his inventions) "he did only start and design, and that they have since been carried to perfection by the industry of others, yet it is reasonable that the original invention should be ascribed to the true author"—Christopher Wren.

By nature Wren was shy and retiring: a delicate boy, he became a man of small stature.

Who, today, would ask the professor of anatomy at Oxford to design a great university building? Yet Wren cheerfully, and confidently—and most successfully—designed the chapel for Pembroke College, Cambridge, and the Sheldonian Theatre at Oxford, before he undertook his life's work upon the new St. Paul's and the restored City of London.

Today, I am sorry to say, that could not possibly happen. Committees, professional associations, planning authorities, trade unions, all would have something to say. Everyone now seems to be in the straight-jacket of his own narrow job. Unless you have 'trained' got a few letters after your name, and learnt the technique of something or other in an organized way and at the right places, you cannot hope (and few people try) to do anything else. We have no amateurs: indeed the word has become almost a word of contempt.

What a pity! Boys and girls who want to write, act, broadcast, produce films or plays, are far too inclined to wait until someone comes along to teach them how to do it and, after some examinations, present them with a piece of paper saying pompously that they now can. It would be so much better if (like Sir Christopher Wren) they got on with it, if they just started to write or act, and took every opportunity they could to *do* what they want to do. Of course teachers and instructors can help but you teach yourself most by practical experience, by getting on with your chosen interest.

Wren could draw; he knew much of mechanics and something of engineering. He had traveled a little and seen the palaces and churches of Paris but I doubt if, before he was thirty, he had ever designed anything much more ambitious than a telescope. What admirable courage, then, what a superb belief in himself, Wren displayed when he suddenly turned his hand to designing great public buildings and the replanning of a whole capital city!

He was the Great Amateur

V

The building of Wren's masterpiece, St. Paul's, was not begun until 1675. The labour went on during the 1680s and the 1690s. By 1697 the choir and the lower parts of the dome structure were completed, though the form of the west end and of the high visible part of the dome had not yet been finally determined. No wonder Wren lived to such a great age; it was almost necessary that he should.

As an old man, hobbling around London on a stick, fifty or more years after the Great Fire, he might feel that he had been forced to compromise, to make the best of a bad job. London was not his London, not the London he and King Charles II had hoped to see rising from the ashes of the Fire. But the spires of his many churches now pointed to the sky, dominating the City in a way we cannot now conceive, because the high cliffs of office-buildings dwarf them and hem them in. His great cathedral, pomp in stone, crowned all.

Wren had more than good reason to feel a sense of achievement : his lasting monuments were everywhere: he had only to look about him.

Count Zeppelin

(1838–1917)

Airship Dreamer

No invention in history has made such a dramatic come-back, after staying in obscurity for more than 40 years, as the *Zeppelin* or airship. With the availability of new materials, this once-upon-a-time popular mode of transport will soon appear in the skies of many western countries. Experts agree that the reappearance of these mammoth whales of the air is likely to herald a new era in transport. At last, the dreams of their long forgotten inventor Count Ferdinand von Zeppelin will be realised.

Count Zeppelin was forgotten because his invention failed to attract the masses. Not much is therefore known about his personal life. He was born on July 8, 1838, at Konstanz, now in West Germany, in a noble family. Boys of noble families in those days sought a military career. Count Zeppelin also received a military education and became a cavalry officer at the age of 19. He took an active part in the Central European wars of 1866 and 1870-71. In millitary circles he had made a name as a daring long-range reconnaisance scout and bold fighter when on patrol duty. When he was 25, he went to the U.S.A. to watch the American Civil war as an observer on the side of the Northern States. In St. Paul, Minnesota, he saw for the first time a hot air balloon which was then being used for military reconnaisance. Later, in Canada, he even had the privilege of boarding the gondola of a hot air balloon and going up in the air.

In those days, a hot-air balloon was simply a huge silk balloon filled with hot air. A gondola to carry passengers was attached to the mouth of the balloon. Once up in the air, the balloon drifted with the wind. The first trip in a hot-air balloon left a deep impression on Count Zeppelin. The dependence of the movement of the balloon on the wind irritated him. Another disadvantage he noticed was that if, by any chance, the balloon was damaged or developed a leak, it would then fall down. So when after nearly a decade he conceived of a balloon that could fly, it had none of these defects.

First of all, instead of flexible silk balloon, Count Zeppelin went in for a balloon of rigid structure. He conceived of a huge cylindrical frame made of light metal aluminium containing small compartments. Each of these was to be filled with bags containing the lightest gas — hydrogen. The advantage of such an arrangement was that even if one or more bags leaked, the overall performance of the balloon was not affected. The entire cylindrical frame was then to be covered with a fabric. Below the balloon were to be two cabins for passengers. To direct its movement through the air were two diesel motordriven propellers and a rudder at its tail end. Count Zeppelin wanted to drive his balloon or airship in the air like a steamer through the seas.

Count Zeppelin was able to convert his idea into a real airship only at the age of 62 after he had resigned from millitary service. In the beginning he sought help from the military authorities to finance his project of building airships. However, when the authorities studied his ideas, they thought that the old Count had gone out of his mind. The Count therefore started his own firm to build airships. Although his friends gave some money to start the firm, the major share was his. In a small shed near Lake Constance, Friedrichshafen, he began the construction of a huge airship. In due course, he realised that the shape of the balloon should not be cylindrical but should be streamlined. Therefore his airships had a cigar shape. The weakest point, which was the main cause of failures, was the motors that drove the propellers. The available motors were not powerful as well as lightweight. It was only after he passed away that lightweight motors of enough power were built and installed in airships.

Finally on July 2, 1900, an airship — 140 metres long and 13 metres in diameter — rose up and moved through the air breaking all previous records. This event took place only three years before the Wright brothers made a successfull flight in their airplane Kitty Hawk. Naturally, when people saw the whale-like airships flying quietly over their head there was a surge of admiration for the old Count. The airships were therefore lovingly referred to as "Zeppelins". But things did not turn out so easy for the Count. Besides achieving success in flying his airships for as long as 80 minutes, most of them also met with accidents, which eventually left him penniless. When he appealed for help to his countrymen, the people gave him an overwhelming response. Within a few weeks six million gold marks were collected and a Foundation for building airships was formed !

World War I began and came the period of trial for airships. Besides being used to keep watch on submarines, airships were used to make raids on London. Some 40 airships were reported to have been destroyed during such raids. Their large size and slow speed made them easy targets for enemy gunners. In 1917, the Count passed away sure that his airships could not win the war for Germany. But that was not the end of airships. Other countries like UK, France and Italy had also begun to build airships. However, when a large number of airships met with accidents one after another primarily due to the highly inflammable hydrogen that filled them, the public began to lose faith in airship travel.

Finally, when a giant airship *Hindenburg* blew up over New Jersey, USA, in 1937, killing its 56 passengers the fate of airships was sealed once and for all, although one German-made airship the *Graf Zeppelin* (Graf means Count) had carried several thousands of passengers and several tons of freight across the Atlantic more than 100 times and had once even been round the world without a single mishap. However, the Wright brother's airplane had now arrived on the scene and had made air travel much faster and immune to bad weather.

One cannot, of course, say that airships were totally out of use all these years, thanks to a few enthusiasts who built them off and on for advertising purposes. The recent interest in airships had been revived mainly due to the efforts of these enthusiasts. Now the highly inflammable hydrogen can easily be replaced by the inert helium, the second lightest gas, which can be obtained in large quantities. Huge airships — as long as 500 metres — can also be built from the now easily available lightweight but tough materials such as new alloys, plastics. fibres, which can increase the load-carrying capacity of the airships. Besides, airships have an advantage over aeroplanes in that no large aerodromes are required.

Airships can therefore be used for carrying men and goods to inaccessible regions such as deserts, mountains, dense forests, and where no rail or roads are available. The prevailing energy crisis further guarantees the return of airships.

Helitruck For Villages

Suppose some goods have to be delivered to a village in a jungle. Then a railway wagon would have to unload the goods at the nearest railway station, from where a truck would carry it to the town nearest to the village. From the town, a bullock cart would carry the goods right upto the village, provided the roads are not very bad. Apart from the fact that the goods would take a lot of time to reach the village, they would be expensive too because they have been brought from a distant place. Moreover, only those goods which would not rot during the long journey would be transported. As a result, villagers are deprived of so many items that city dwellers enjoy.

Can there be some means of transport which could quickly deliver goods to a village at a low price? The German Agency for Technical Cooperation intends to manufacture a means of transport for the villages of the world. It is called "Helitruck", a kind of balloon aeroplane and helicopter combined. It is a huge cylindrically shaped balloon containing the second lightest gas, Helium. Like a balloon, it would rise up in the air. Four propellers would further enable it to take off vertically like helicopter, and also move like an aeroplane once airborne. Travelling at a speed of 20 kilometres per hour, it can easily cover 4,000 kilometres at one go. It can again land like a helicopter on a small stretch of plain normally available in any village. Helitruck would especially be of immense use in a country like India, where roads to interior villages have not yet been laid or are too expensive to lay. Helitruck would certainly open the gates of prosperity to villages, if it is brought into use in India.

Don Bateman

Saviour of The Skies

Don Bateman was poring over a cockpit-wiring diagram when a colleague walked into his office one day in 1966. "You might be interested in this," the man said, handing him a single sheet of paper. The trim, 34-year-old engineer scanned the document issued by Scandinavian Airlines (SAS). On a wet night a few years earlier one of its airliners had crashed while trying to land at Ankara in Turkey, killing all 42 on board. Now the company wanted someone to invent something like a fire-bell to warn pilots when they got too close to the ground.

"I'd be really interested in having a crack at this", Bateman said. His colleague wasn't surprised. Most of the people at United Control Inc, a small US aerospace electronics firm in Washington, knew about his passion for safety.

Bateman had long been haunted by an episode from his childhood. One afternoon, just as the bell rang to end his school day, two training planes collided in mid-air over his hometown in Canada. As wreckage rained over the neighbourhood, the seven-year-old pedalled his tricycle furiously to the crash. He'd never forgotten the sight of the crew-members' broken bodies.

The accident hadn't quenched the youngster's burning desire to become a fighter pilot, but a diagnosis of latent diabetes ended his dreams of an air force career. Bateman, who'd built ham radios all his life, became an engineer instead.

Jobs were scarce and America's burgeoning aviation industry lured him to that country.

Now SAS's request triggered a novel idea. Bateman took data from two cockpit warning devices. One alerted pilots when their plane was in danger of a stall. The other, an altimeter, bounced a radio beam off the ground to measure a plane's precise height so it could be landed in fog.

Bateman incorporated the data into a new device which lit a red light in the cockpit when the ground was rising to meet the plane, giving a pilot time to climb out of trouble. Later, he made it sound a loud *Whoop-whoop!* Bateman called it the ground proximity warning system (GPWS) or "ground-prox."

After three years of discussions *SAS* began fitting their aircraft with ground-prox in 1972. Pilots loved Bateman's $5000 system, but other airlines showed little enthusiasm. For Bateman the pace was frustratingly slow.

America's Federal Aviation Administration (FAA) wasn't interested, either. This kind of accident, technically called "controlled flight into terrain" (CFIT) because the aircraft was in control and operating perfectly until it crashed into the ground, usually happened when pilots lost their bearings in fog or darkness or were distracted. It was easy to blame them on pilot error and the FAA argued that it was the pilots' duty to stay out of trouble.

But Bateman was convinced that ground-prox had tremendous life-saving potential. While mid-air collisions have frightened people the most, far more planes inadvertently flew into the ground. Bateman figured that all 3500 lives lost over five years could have been saved by ground-prox.

Bateman continually refined his device, tested it in the air and showed how it would have saved lives. One day in September 1971 his company plane, fitted with ground-prox, flew towards the Chilkat mountain range in Alaska. Straight ahead, the confetti spilling from a crater was all that remained of a Boeing 727 that smashed into the mountain in fog a few days before, killing 111 crew and passengers.

Bateman was retracing the fatal route and, sure enough, seconds before reaching the rock face the ground-prox alert sounded, giving the pilot time to pull up smoothly and zoom over the ridge.

"We've got to get this box into every plane," Bateman said. He continued demonstrating the ground-prox to pilots and safety experts, and kept on improving the device to give pilots a few seconds more warning time.

Still, another three years passed until December 1, 1974, when a TWA 727 clipped a ridge some 30 kilometres from Washington's Dulles airport. All 92 people on board perished— and many of them had lived and worked in America's capital.

There was all outcry, and Bateman knew it might be now or never. He sent the company plane across the United States, repeatedly retracing the track of the 727 to the impact crater, showing politicians, safety experts and the media how ground-prox would have alerted its pilots 12 seconds before the crash. The victims need not have died.

This time the message got through. The *FAA* made ground-prox compulsory in all commercial jets within days.

The effect was dramatic. CFIT accidents in the US fell from an average of eight aircraft per year to only one aircraft every five years.

As timely warnings saved more and more lives, the modest Bateman became known as "Mr Ground-prox." His system was now basic equipment the world over, his place in aviation history secure. Most men would have been content to rest. But for Don Bateman it was only the beginning.

As the years passed, Bateman got more and more complaints from the airlines that the radio beams bounced off other planes, even hail, and caused false alarms. This lead pilots to ignore the warnings with deadly consequences.

In one case Avianca 747 was descending into Madrid when the ground-prox sounded an alert. "Okay, okay", the captain muttered, doing nothing. Seconds later the plane piled into a ridge eight kilometers short of the runway, killing 183.

Every accident hit Bateman hard. He'd read aloud from the newspapers the names of the crash victims. "Dammit, we should have saved these people!" he told his fellow engineers.

With grown-up children and a lakeside cabin, Bateman was approaching 60, the age when most men began coasting towards retirement. But Bateman still ran several marathons a year, and during long training runs along woodland trails he thought about how to make his invention even better.

A cruise missile, he knew, "read" its way over the terrain using a radio altimeter linked to the route map in its computer memory. Bateman wondered if an aircraft could do the same sort of thing—but using precise, Global Positioning Satellite, (GPS) navigation systems instead of radio beams to fix its position. The device Bateman had in mind would require vast amounts of computer memory and very fast processors. But microchips were getting better every year.

Finally, one day in 1991, Bateman called some of his most trusted engineers into a back office and shut the door. "Let's get going on this thing," he said.

BANKING in the sunlight, the small, twin-prop plane left an airfield leaded for the ice-scoured Cascade Mountains. Bateman indicated a craggy peak and told pilot Markus Johnson, "Fly at that one."

Cradling a laptop computer, engineer Kevin Conner grinned triumphantly as he showed Bateman the picture appearing on the screen — a moving contour map of what they could see out of the window. The "enhanced" ground-prox (E-GPWS) was born. The computer was reading the plane's precise position several times a second : from the GPS. Meanwhile software compared this position to a digital contour map looking ahead to "see" the terrain kilometres away.

The device had tremendous potential, but came at the worst possible time. Bateman's company was being groomed for sale and the management was cutting costs to the bone The situation didn't get better when the company became a division of AlliedSignal (later Honeywell).

Bateman went to see his superior, Frank Daly and snapping a video cassette into the office player said, "Take a look at this Daly saw the new ground-prox operating in the cockpit of the company plane. He was impressed.

"I don't need a lot of money", Bateman said "I just want to be left alone". Daly agreed.

But the company's financial squeeze grew tighter. A few months later Daly told Bateman he'd have to shut down for a while.

Bateman said little but his mind was racing. By now he'd painstakingly investigated and put together reports of more than 300 CFIT crashes. Pulling out some recent ones, he compiled a report showing that some 2000 passengers would still be alive had their planes been fitted with the new ground-prox. He put the report on Daly's desk.

"We're not doing this for money," Bateman said with quiet passion. "We're doing it because we must."

"You win," Daly said, and hid Bateman's cost in the budget as factory equipment.

Now Bateman drove his team hard and himself harder. There was endless fine tuning, endless testing, but when energies flagged, Bateman was unyielding.

Once he called the whole team into his office and played an audio-tape. It was recorded in the cockpit of a Yugoslavian DC-9 charter jet trying to land at Ajaccio, Corsica. Suddenly the old-style ground-prox sounded an alert, but the pilot didn't pull up in time. There was a bang when the wing clipped a rock. As the jet turned on its back and began to plunge, the pilot's young son screamed, "Daddy, I don't want to die!"

"Don't lose sight of why we're doing this stuff," Bateman told them gruffly.

His team soon ran into a huge problem: the digitized terrain maps that were the heart of the new system existed in only a few places. Batman put engineers Frank Brem and Christine Stahl on the job. "We'll have to do this ourselves," he said, "and I want data for the whole world."

The US and Canada released the data without a problem, but other countries were harder to crack. France was ready to sell terrain data of the whole country for $ 35,000; Britain demanded $ 500,000 plus a royalty from every aircraft using it. Most of South America and Asia regarded terrain data as a military secret; others, especially in Africa, just didn't have it.

Brem scoured the world; Bateman himself rummaged through maps of the Himalayas in London's British Museum. He even found a complete set of World War II maps of Turkey in the University of Washington archives and had them digitized.

Ultimately several agencies, some official and some not, sold them a superb set of maps compiled by the former Soviet Union.

By 1994, enhanced ground-prox was ready: three circuit boards slotted into a box the size pf a telephone directory with a map of the world stored on a chip. At $ 55,000 it was cheaper than a new coat of paint on a jet.

And then, astonishingly, Bateman hit the same brick wall as 20 years earlier. Senior pilots thought it was vital; the airlines didn't, and neither did the FAA.

Dismayed, Bateman went to London and met David Fleming, chief technical pilot of British Airways (BA). Fleming put the device in a 747 and flew it with Jock Reid, chief test pilot of the UK's Civil Aviation Authority.

Even from the end of the runway at Heathrow, Fleming and Reid saw Windsor Castle on its hill, 13 kilometres ahead, displayed on the ground-prox screen. Then Reid flew the plane straight at the 1085-metre peak of Mount Snowdon, in Wales.

A full 90 seconds before he would have crashed, ground-prox sounded "Caution, terrain!" "Outstanding," Reid said.

Bateman's company was now behind the project foursquare and they had started discussions with American Airlines, but once again progress was frustratingly slow.

Then a few days before Christmas 1995 an American Airlines (AA) 757 crashed in low clouds on approach to Cali, Colombia, killing 159 people. From his contacts, Bateman learned the old style ground-prox had sounded 12 seconds before impact. The pilots pulled up, but the warning had come too late and the aircraft flew straight into a mountain. The new ground prox would have saved them.

Three days later Bateman was in AA's Dallas headquarters, addressing a room packed with senior pilots and safety managers. Hooking up a laptop and projecting the image on a wall screen, he simulated the doomed jet's flight path. The outline of the mountain appeared in bright red.

The company did their own testing and several months later began placing orders.

Alaska Airlines started to equip their fleet, and even the FAA was convinced. All passenger planes in the US must have enhanced ground-prox by 2005. As for India, the deadline is January 1st, 2003.

Improvements to enhanced ground-prox are being made; transmission lines and radio towers are being programmed into digital maps, for example, so helicopters can use ground-prox. But whenever progress comes too slowly, Don Bateman, now 68, still pushes hard.

"Don't tell me it can't be done!" he'll say climbing on a conference table and stamping his foot. "Every day of delay means more lives lost. Let's get on with it!"

How Ground-Prox works

This is the display that the pilot sees if the aircrafts is in immediate danger. The solid red area represents ground which the plane risks hitting within about 30 seconds or less. The solid yellow area indicates dangerous terrain within about 30-60 seconds flying time. Green dots represent land that is below the aircraft. Yellow dots mark terrain that is at approximately the same altitude as the airplane. Red dots show ground that is well above the aircraft. The open white triangle at the bottom of the photo represents the position of the aircraft.

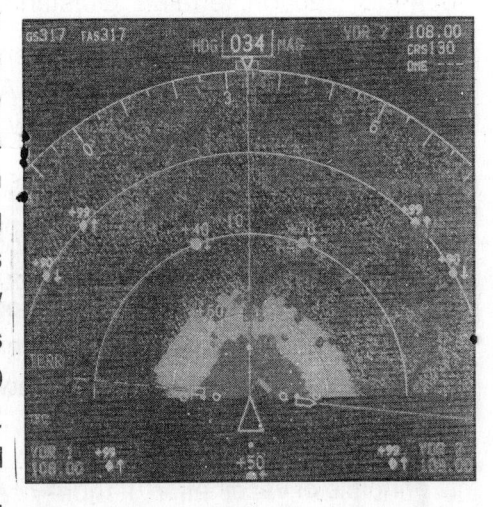

Edward Jenner

(1749–1823)

Acute observation, determination, long years of patient research and experiment, and courage: these, and the brain of Edward Jenner, fought and defeated the scourge of small-pox. Basing his work on the country belief that infection with cow-pox prevented the contraction of small-pox, Jenner worked for twenty years to establish at last his principle of vaccination. It rapidly came to be recognized as an effective preventive of small-pox, and its widespread use has practically stamped out this once rampant disease.

> The sun drove off the twilight gray
> And promised all a cloudless day;
> His yellow beams danced o'er the dews
> And changed to gems their pearly hues.
> The song-birds met on every spray,
> And sang as if they knew the day;
> The blackbird piped his mellow note,
> The goldfinch strained his downy throat,
> To join the music of the plain;
> The lark pour'd down no common strain;
> The little wren, too, left her nest
> And, striving, sang her very best;
> The robin wisely kept away,
> His song too plaintive for the day.
> 'Twas Berkeley Fair, and Nature's smile

> Spread joy around for many a mile.
> The rosy milkmaid quits her pail,
> The thresher now puts by his flail;
> His fleecy charge and hazel crook
> By the rude shepherd are forsook;
> the woodman, too, the day to keep,
> Leaves Echo undisturbed in sleep;
> Labour is o'er—his rugged chain
> Lies resting on the grassy plain.

These lines were written by a young Englishman living in the county of Gloucester, in the last quarter of the eighteenth century. His name was Edward Jenner, and he had spent all his life in his father's vicarage at Berkeley, where he was born on May 17, 1749.

Surrounded by the lovely Cotswolds, in daily contact with farmers and men to whom industrialism was as yet unknown, Jenner was steeped in a love of the countryside in general and his native Gloucestershire in particular. As a boy he could recognize the cry of every bird, and he could name every plant by the roadside. To him, therefore, observation came naturally, and it was not long before he began to turn his knowledge to good account. He so loved the beauties of nature that to him disease and sickness were as sacrileges against his beloved goddess, and he determined to become a doctor.

Poetry was abandoned, and he began his studies under Dr. Daniel Ludlow, a surgeon, of Sodbury, near Bristol. It was during this time that he noticed a local belief that if once a man had had the disease known as cow-pox, usually caught from sores on cows' udders when milking, he could not contract small-pox. Knowing that there is no country saying which has not some foundation in fact, Jenner determined to see what truth there was in this, and he made ceaseless enquiries among his medical colleagues to get to the bottom of the matter. But he did not receive much help. Most doctors seemed to think that this popular notion was just an old wives' tale and nothing else. Still Jenner was not satisfied and, after he had gone to London to study under the celebrated physician, John Hunter, he kept on with his enquiries and investigations, when he returned to begin practice at Berkeley in 1773.

Finally, in 1780, he discovered that there were in reality two different forms of cow-pox, only one of which acted as a defence against small-pox. At that particular period the cow-pox happened to be scarce in Gloucestershire, and some time passed before he could test his theory. But, in the meantime, Jenner collected every scrap of information he could, and published each step of his discoveries.

At last, after more than twenty years of research, his theory seemed to him to be irresistibly proved, and he resolved to carry out his first experiment of inoculation. On May 4, 1796, he inoculated an eight-year-old boy whose name, James Phipps, should be blazoned in the annals of medical history, with cow-pox vesicles from the hands of a dairymaid, named Sarah Nelmes.

On July 1 he inoculated the boy with "variolous matter," or small-pox germs. His medical colleagues awaited the result with ill-concealed excitement, not a few criticizing Jenner in the harshest and most unscrupulous manner for the risk he was taking. But Jenner was immeasurably above such criticism. He knew that the conclusions he had reached were, by all the standards of medical science as they were then known, unassailable; and finally, as he had predicted, James Phipps escaped the small-pox. As is the way with the small-minded, his colleagues turned from criticism to immoderate adulation, and it was not long before they advocated vaccination for almost every disease to which the human frame is liable.

Joseph Farington records in his famous diary: "Before he (Jenner) published his discovery, Sir Walter Farquhar said to him that if he chose to preserve it a secret, he might make £100,000 by it. It would be easy for him to prove its value to medical men of character who would recommend it and warrant its efficiency, which would enable him to get £10,000 a year by it; but Dr. Jenner determined to give it to the world."

And he was as good as his word. In 1798 Jenner published his *Inquiry into Cause and Effect of the Variole Vaccine* which be followed a year later with *Further Inquiries* and, in 1800, *Complete Statement of Facts and Observations*.

By 1806 the great social reformer, Samuel Wilberforce, was able to say that "even in remote countries, and even in China, a country in which innovation is jealously opposed, it has been admitted. In India it is used."

Like all great men, Jenner was essentially modest by nature and yet impervious to unscrupulous opposition. He had two kinds of opposition to face. The first was the direct attack of the conservative medical men of the time, who regarded vaccination as a dangerous practice; the second was much more difficult to deal with.

No sooner had he published his results than every kind of doctor, both genuine and quack, took up the practice and advocated its adoption. The village fair-grounds were infested with quacks offering "the genuine vaccine operation of Dr. Jenner," and the people that were thus infected with diseases were innumerable.

One of the most notorious of the genuine medical men who advocated vaccination, while being completely ignorant of its real nature, was a certain Dr. George Pearson. This man published a pamphlet in November, 1798, in which he gave to a troubled world his theories on the subject without having even seen a case of cow-pox. He delivered frequent lectures on the subject, and inoculated hundreds with a virus which was found to produce not cow-pox, the slight disease produced by Jenner, but more or less severe eruptions very similar to small-pox.

Jenner at once proceeded to look into the sources whence Pearson had got his vaccine, and he found that, as he had suspected, this vaccine had come into contact with variolous matter. But this did not deter the amazing effrontery of Pearson.

Hearing that Jenner was trying to found institutions where people could be vaccinated free, Pearson proposed founding such an establishment in London, and actually had the effrontery to offer Jenner the post of honorary consulting physician. Jenner immediately went to London to counter this move. The insult to himself did not disturb him, but what did cause him the most acute anxiety was the thought that an institution was to be formed where erroneous and dangerous practices would be carried on. There were, fortunately, at that time a few men of influence in London who recognized Jenner's worth, and George Pearson's scheme was sent to the obscurity it deserved.

Other and even more unscrupulous attacks were made against Jenner. A particularly vile calumny was the story that was spread that Jenner had given up vaccination, and was inoculating people with a mild form of small-pox. Colour was lent to this libel by the fact that Jenner had had his younger son, Robert, inoculated with the small-pox virus; and the tale went the round of England.

The real truth was that Jenner had vaccinated his son, and that the vaccine had not "taken" — that is to say that the expected mild infection of cow-pox had not taken place. But one day, a friend of the family called on the Jenners when they were at Cheltenham, and began to play with young Robert. In the course of conversation he remarked casually that he had just left a family where small-pox was raging. This plunged Jenner into a fever of anxiety and he exclaimed, "Sir, you know not what you are doing-that child is not protected. He was vaccinated but the 'infection failed'."

As he was away from home, Jenner had no vaccine with him. So he did the next best thing, which was to have the child inoculated with small-pox, since he knew that the illness brought on by inoculation was only a very mild form of the disease. This action was perfectly natural, and in the circumstances the only proper thing for Jenner to do.

But still the story ran its round, until it came back to Jenner at a reception at St. James's, where he overheard a certain peer, who was personally unknown to him, repeating it as a tit-bit of gossip. Jenner's methods were always direct, and in none of his actions did he ever lose possession of his calm dignity. Going up to the nobleman, he said, "Sir, I am Dr. Jenner." That killed the spread of the calumny as far as London was concerned, and, little by little, the real facts became known.

In the meantime, Jenner's fame spread far beyond the shores of Britain. In France vaccination was widely practised, and Napoleon became an enthusiastic patron of the institution set up for its spread. On hearing of Napoleon's advocacy of his discovery, Jenner wrote to him on one occasion to petition for the release of some English prisoners. The letter was forwarded to the Empress Josephine, and she went to plead the cause of the imprisoned Englishmen. At first the Emperor was adamant, but when Josephine said that it was Jenner who was pleading, Napoleon said, "Ah! we can refuse nothing to that name."

So highly was the name of Jenner thought of that certificates signed by him acted as passports and secured the release of other prisoners in countries as far apart as Mexico and Austria. Those who ruled in England at the time were less large-minded, and Jenner's efforts for the release of certain French prisoners were unsuccessful.

Parliament did, however, vote him £10,000 towards spreading his discovery, but the money was not paid to him until two years after the vote, and then £1,000 was deducted in "government fees." The sum Jenner thus received was inadequate, and did little more than pay the expenses he had incurred in making his discovery. Some idea of the extent to which Jenner had to put his hand into his own pocket may be gained from the fact that he once wrote to a friend describing himself as "the vaccine clerk of the world," and Farington recorded in his diary in 1806 that Jenner's enormous correspondence alone cost him £100 a year.

But money never attracted Jenner, and he was as ready to vaccinate the poor as the rich. He offered free vaccination to those who were so poor that they could not afford a fee, and as many as three hundred of the poorest would wait at his door every day to be vaccinated.

Parliament was finally prevailed on, in 1806, to vote him a further £20,000, this time without any deductions, and with this money Jenner founded, in 1808, the National Vaccine Institution. He spent some months in London organizing the institution, and then one of his sons fell dangerously ill, and Jenner returned to Berkeley.

While he was away, Sir Lucas Pepys, President of the College of Physicians, formed a board composed entirely of that college and the College of Surgeons, in flat contradiction of Jenner's orders, and so Jenner immediately resigned his post as director of the institution.

A short while later, Oxford conferred on him the degree of M.D., and everyone believed that this naturally meant election to the College of Physicians. But that body refused to elect Jenner unless he passed an examination in classics. Jenner's refusal to brush up his Latin and Greek was characteristic. He said, "It would be irksome beyond measure. I would not do it for a diadem. That indeed would be a bauble: I would not do it for John Hunter's museum."

In 1810 his eldest son died, and this loss and the strain of his labours caused Jenner to have a breakdown. He retired to Berkeley, only going up to London on important business.

His last public appearance in London was in 1814, when he was presented to the allied sovereigns, who were in London following the Congress of Vienna and the banishment of Napoleon to Elba. Jenner then went back to Berkeley, where he had lived in retirement, enjoying, uninterruptedly for only too brief a period, the company of his wife, to whom he was devoted.

He had married, twenty-six years previously, Katherine Kingscote, the daughter of a Gloucestershire gentleman, and whenever he was able to snatch time from his many visits to London in connection with his discovery, he loved to take his wife and young family to Cheltenham for a holiday. He had a high opinion of the Cheltenham waters and recommended them to all his patients.

His biographer and friend, Dr. Baron, says that "his wife was gentle and mild, and exercised a great and peaceful influence on him ... Jenner was always conscious of the presence of God, and Mrs. Jenner taught this lesson to the poor around her in schools she had set up for teaching the Scriptures." But Mrs. Jenner was always delicate, and this caused her husband constant anxiety. When she died, in 1815, Jenner's grief was overwhelming. He wrote to Baron shortly afterwards: "Every surrounding object reminds me of my irreparable loss. Every tree, shrub, flower seems to speak. But yet no place on earth would at present suit me but this ... the bitter cup has a kind of relish in it which it could afford nowhere else."

Though Jenner had left London for good in 1814, he was not idle at Berkeley. He devoted himself to his boyhood pursuits of natural history and geology and he kept well abreast of all the tendencies of his time. He built the first balloon that was ever seen in Gloucestershire, and he did much to further the geological knowledge of the county. In 1823 he presented his last paper to the Royal Society—of which he had been made a Fellow in 1780—"On the Migration of Birds."

But although to all appearances he seemed destined to live many years longer, his unflagging industry and care for others had taken its toll of his strength, and he was found on the morning of January 24, 1823, unconscious on the floor of his library. He had been stricken with apoplexy. His right side was completely paralysed, and it was clear that there was no hope. The next day this simple-hearted lover of Nature, poet and 'great' benefactor of mankind passed to his reward.

Eli Whitney

(1765–1825)

No invention was made within two weeks after the inventor was requested to try his hand at it. No invention proved to be so valuable to users and so useless to its inventor as this was. No invention changed the landscape of country as this one. What was this invention?

It was the invention of cotton gin — an engine for separating fibres from seeds of a cotton plant. Its inventor was Eli Whitney. The then American President Thomas Jefferson called him " the Artist of the country" because he changed America into a rich and prosperous country. Besides, he is also renowned for inventing what is today known as "mass production method" which Henry Ford, the inventor of car, employed successfully in his factory at the beginning of the last century.

Eli Whitney was born to prosperous farmer of Westborough, Massachussetts, U.S.A., on December 8, 1765, when that country was in political turmoil. He was curious from his childhood. He used to regularly visit the workshop in his father's farm and worked on a lathe machine to make or repair household things and agriculture equipment. Once when his father said that his watch was running slow, he opened it in his absence but could not reset it. Afraid of his father's anger he feigned illness, did not go to school, and somehow reset the watch. At the age of 12, he built a violin which had tolerably good musical quality. During dances in the neighbouring farms, he used to play it and earn some pocket money. Unfortunately, at this age, his mother passed away and two years later his father married again. His stepmother never liked him. Owing to tensions at home, young Whitney decided to be on his own. He asked his father to set up for him a forge in the workshop. He thus began to manufacture nails, which in those days were in short supply in the country due to the war with England. He was highly successful as long as the war continued. When foreign supply of nails was resumed, he switched over to making pins for ladies' hats and walking sticks. Without any guidance, he developed a technique of drawing steel into long and fine pins.

At the age of 17, when most boys gave up education in those days, Whitney suddenly felt a strong desire to educate himself. He knew he could not depend on his father. He therefore decided to become a school teacher, so that he could earn while he learned. Sounds funny today — isn't it? — because Whitney himself had no education. In those days, as no one was properly educated, town authorities used to select a school teacher on basis of his intellect, and not degrees. Thus Whitney became a teacher in a school. He used to study at night to keep ahead of his students. In due course, he saved enough money so that he could join an academy at Leicester during summer holidays. In three years' time he learned enough arithmetic, Latin and Greek to get admission into the then prestigious Yale College. His father was impressed. Against his stepmother's objections, he gave him enough money to pursue his studies at the college. Here

Whitney not only improved his personality but also made a number of friends. After completing the college course, he wanted to take up law. But, like his other classmates, he knew he could not afford that luxury. So, at the age of 27, Whitney found himself educated yet penniless young man.

When Whitney did not secure any good job, he had nothing but to opt for the tution secured by a former Yale colleague, Phineas Miller, in the South of the country. Miller was the manager of Mulberry Grove, Georgia, the estate of the late famous General Nathanael Greene. Along with Catherine Greene, the charming widow of General Greene, and her two children, he was going south to their estate. He therefore asked Whitney to give them his company. Unfortunately, the journey proved to be a series of disasters. On the way, Whitney became sick, was almost drowned when the ship wrecked, and caught smallpox. To top it all, he came to know that his tution charges had been reduced by half when he reached Georgia. Seeing his miserable situation, Mrs. Greene invited him to stay at Mulberry Grove as a guest. Within a few days, he repaired some toys of her children and built an embroidery frame for needlework. She was impressed. it was when some cotton planters from the neighbourhood came to seek her advice that on the spur of the moment she called Whitney and introduced him as the man who could solve their problem. He was called as the man who could work wonders. And so he turned out to be.

Whitney could not understand what to do because he had not even examined a cotton plant carefully. How could he solve a problem concerning it? He shrugged his shoulders and promised to look into the matter because he had nothing else to do. In those days, two varieties of cotton plant were grown in the U.S.A. One cotton variety had black seeds with long fibres. Using a mechanical device, like the cloth-wringer — two cylinderical rollers rotating in opposite directions — seeds were easy to separate from fibres. However, the disadvantage of this variety was that it could only be grown in the coastal regions of the country. On the other hand, the other cotton variety — green seeds with short fibres — could be grown all over the country. It had however one major problem. When fed into the machine, seeds were crushed and scattered throughout the fibre. It was extremely difficult to remove seeds from fibres. It took a person one full day to process one pound of cotton. No machine had yet been invented which could perform this task of separating green seeds from fibres. However, in anticipation of the invention of such a machine any time, some planters had already grown green seeds in 1793. But when the harvest time came, no such machine had been invented. Puzzled, the planters thought they would seek the advice of the widow of their old commanding officer, who was always ready to offer any kind of help. It was this problem of separating seeds from fibres of green cotton plant variety that had been presented to Whitney.

Miller also encouraged Whitney to try his hand and provided him all facilities for the invention. A room was evacuated in the farm house for Whitney's workshop. A parcel of cotton and all the other raw materials and tools were made available to him in no time. Whitney drew a sketch of the machine for processing cotton. By the tenth day, he had built a little working model of it. In due course, a full-scale model of the machine, which he called "Cotton gin", where gin is a short name for engine, was built. In the cloth-wringer type of machine, seeds were separated from fibre. Whitney thought of the reverse way: separate fibres from seeds. His cotton gin consisted of a huge cylinder, 15 cm in diameter and 65 cm long, which had teeth made of coiled iron wire. The teeth, when the cylinder was rotated, passed through a breastwork — narrow slits made of metal— which allowed only fibres to pass through but not seeds or crushed seeds. Fibres were literally torn from seeds as they passed through the breastwork and were flung out to be collected. The seeds meanwhile were thrown in the opposite direction and were collected at that end.

It is said that when Whitney demonstrated the working of the cotton gin for the first time to Miller and Catherine Greene, the gin stopped working after some time. The cylinder could not rotate as the fibres had clogged its teeth. Whitney was puzzled but

Catherine Greene immediately suggested him to add another cylinder with brushes, which could be rotated in the direction opposite to the main one so that it could clean up its teeth at the same time. She gave him a brush which was used in her house to clean up fire hearths. Indeed, the addition of the new cleaner cylinder led to a smooth functioning of the gin. It could process cotton ten times faster than a man could and process it cleaner and better. A man or horse could be used to drive it. In fact, one man and one horse could do the work which fifty men could do with the old machine. One can easily imagine from the above description how simple in design and working is the cotton gin. Any body who had once seen its working could re-invent it taking the assistance of a carpenter or iron smith. And that is what actually happened.

In a short time, Catherine Greene annuounced a public demonstration of Whitney's cotton gin at Mulberry Grove. Planters from the neighbourhood came to see the demonstration and admired the genius of Whitney for solving their problem within such a short time. Soon, they began to bring their recent harvest of cotton to Mulberry Grove to get it processed. Demand for the gin-processed cotton also increased because it was cleaner and fetched more money. The demand for the gin was beyond either Whitney's or Miller's or Cathering Greene's expectations. Miller took this opportunity to sign an agreement with Whitney to share profits from the gin. Greene also gave them wholehearted support to start the manufacture of gins on a large scale. Whitney was sent to Philadelphia, where he filed a patent on the cotton gin. He returned to set up a workshop at New Haven and began to build cotton gins on a large scale.

Had Miller and Whitney decided to manufacture cotton gins or given the rights to manufacture them to somebody else, things would not have turned out the way they did. However, they decided to manufacture cotton gins but not to sell them. They decided to install "ginneries", a large number of cotton gins, at certain strategic locations first in Georgia and then in South Carolina. Instead of charging money for processing cotton, they decided to barter; two-thirds of the processed cotton was taken as charges. The reasons for taking these decisions were quite sound. First of all, they were not sure whether any planter would buy cotton gin, and second whether he would pay money for processed cotton. Miller even inserted an advertisement in Georgia Gazette offering their services to neighbouring planters. The first ginnery started functioning at Mulberry Grove and began to process the 1794 cotton harvest. However, the demand for cotton processing was high. The planters had to stand in long queues to get their cotton processed at Mulberry Grove and secondly they were unhappy about the high charges. Some of them therefore decided to build their own cotton gin after seeing its simple construction and working. In a short period a large number of cotton gins began to hum in Georgia. When Miller and Whitney objected, the planters even claimed that they had invented them on their own! One person, Hodgen Holmes, who had seen the original drawings of Whitney, went on to build a modified cotton gin, patented it and began to start manufacturing it. He simply replaced the coiled iron wire teeth by teeth made of iron sheets. Besides, a malicious propaganda against Whitney and Miller started. It was claimed that the gin destroyed cotton fibres. Whitney and Miller had therefore hard time selling their cotton in England.

Helpless, Whitney and Miller began to knock the doors of courts but to no avail. A faulty wording in the patent law helped the copycats. Meanwhile, their factory at New Haven was burnt to ashes destroying 20 almost ready cotton gins. Miller also came under a cloud of suspicion in some other dealing. In course of time, Mulberry Grove was sold and Miller also passed away. But Whitney continued his fight for justice. In 1800 he managed to convince Thomas Jefferson, the president of the country who had earlier granted him the patent for cotton gin, about the faulty wording in the patent law and got it changed. So, as compensation for the losses he had borne, two states, South and North Carolina, paid him a sum of $ 50, 000. Although it was big amount of money, it simply cleared his debts incurred in course of his fight against injustice. In other words, he did not gain anything. In the meantime cotton plantations boomed all over the country. Cotton plants not only fetched cotton for making clothes but their seeds also turned out to be valuable. Oil from the seeds was used in food, soap, grease and fertilizer. Even the shell of the seeds was utilised as a chemical raw material in the production of explosives, shatter-proof glass and writing paper. Had cotton gin not been invented all these things would not have become possible. Yale College therefore awarded him M.A.degree in recognition of his valuable services to the country and the world.

Whitney's invention of cotton gin was purely a matter of chance. Had he not invented it, some one else would have because the demand for it was very high. But the next invention that he made was purely to satisfy a need. In 1798, when bankruptcy and poverty was facing him, he signed a contract with the United States Government. According to the contract, he had to supply 10,000 guns to the government within two years. And, mind it, he did not know anything about guns except that as a child he had once opened and examined his father's gun! It was a sheer act of desperation because till then nobody — not even government factories — had been able to manufacture 20,000 guns in two years. But, then, he received big advance of money, which he promptly passed on to his debtors. It was indeed a leap into the dark. When the government officials asked him how he would manage to achieve his target, he replied that he intended to use water power and machines. As an inventor of cotton gin his word was respected and so the contract was granted to him. Moreover, in anticipation of war with France, the U.S. government was keen that a local individual should manufacture guns. There was none but Whitney who took the challenge.

Whitney was not at all sure of how he was to achieve his target in practice, though he had given a lot of thought to it on paper. In those days, every gun — or for that matter — any product was made by a skilled worker from the beginning to finish. In other words, every gun was built separately. Parts of one therefore could not fit into other gun. If a particular part was damaged, the gun became more or less useless. During a war, this meant heavy loss. Whitney, on the other hand, thought of manufacturing parts of a gun separately. Each part was to be of the same size and characteristics. After all the parts had been manufactured, guns were to be simply assembled. So, even if a particular part of a gun was damaged during a war, another similar part could replace it. The gun could therefore be used again. It was a very attractive proposal for the United Stated government. In fact, this is what is known as the concept of mass production, which is the basis of all industries today.

Moreover, instead of skilled workers, Whitney wanted to build machines which could perform skilled work, an idea he had been harbouring from his youth. He therefore also developed the concept of what is known as "Machine tools", the cornerstone of all industries today.

Whitney basically wanted to take short-cuts to achieve his target but how far he was successful it is difficult to say today. He installed his gun-making factory on a huge stone slab near a fine water-fall in New Haven, so that he could harness water for driving his machines. He also installed moulds and machines. In his later years, the factory became a national symbol and came to be known as "Whitneyville". People from all corners of the country used to flock to the factory to have a look at the machines which manufactured guns. He also gave a part of his manufacturing work to other factories. As he could not manage the factory alone, he employed his nephews for the same. However, despite all his efforts, Whitney could not achieve his target within two years and he had to get his contract extended every year. In all he took ten years to meet his promised target, thanks to the influence he wielded in the U.S. government. By then, he began to feel increasingly isolated so at the late age of 52 he married a lady 20 years younger. Eight years after the marriage, he passed away on January 5, 1825 after suffering from an enlarged prostate gland. After his death, his son took over the arms business when he came of age. All the time that Whitney was suffering hardships, a silent revolution was taking place all over the U.S.A. It eventually raised its ugly head as the Civil War in the history of the country.

More than half the population of the Southern U.S. were negroes, who were employed on farms as slaves. For a long time a feeling was gathering momentum in the country that they had the right to freedom. It was not on humanitarian grounds but on purely economic issues. In the Southern region the land was not fertile and wilderness was in abundance. In course of time, farmers came to realise that they did not have enough work for negroes, who had to be fed in any case. So, to get rid of them, they had begun to accept the idea of their freedom. However, when cotton gin was invented, this feeling took a dramatic turn. Cotton gin enabled farmers in the southern region to produce more cotton. All the wilderness was soon turned into huge cotton plantation which necessarily needed a lot of hands to do labour work. Negroes therefore came in handy. Moreover, a negro, even an old or disabled one, could easily operate cotton gin. In short, farmers developed a vested interest in slavery because without negroes their prosperity was at stake. Thus resulted the Civil War in the U.S.A. Had Whitney not invented cotton gin at the time he did, there would have been no prosperity in the southern region and no Civil War. Negroes would have been set free long time before they eventually were.

Elias Howe

(1819–1867)

Elias Howe was not the first person to think of inventing a sewing machine. Nor was he the first person to build a workable sewing machine. But, certainly he was the first person to invent sewing machine on his own, which was far superior than those built till then. It was also his sewing machine which was first to be patented, modified and manufactured on a large scale. The last link in the chain of machines to manufacture garments after spinning and weaving, it created a revolution in garment industry. In fact, it became a household appliance to be called "a girl's best friend." In the history of inventions, it also became the first machine built to satisfy the needs of women. Even today, it continues to lighten the burden of both men and women at home.

Elias Howe was born on July 9, 1819, at Spencer, Massachusetts, U.S.A., to a well-to-do farmer. He spent his childhood among the machinery at the father's farm. A quarrel with his father forced him to take up a job at an early age. He worked as a weaver and mechanic in several mills. The idea of building a sewing machine occurred to him at the age of 21, when he was already a married man with three children. Owing to his frequent bouts of illness and meagre salary as a mechanic, his wife used to sew dresses late into the night to earn some extra money to maintain the household. Seeing her work day in and day out, he always thought of doing something to ease her burden. There was then always an inner drive in him to invent something novel and make fortune. The idea of inventing a sewing machine struck him when he overheard a conversation between the owner of the shop where he worked and an inventor. The inventor had invented a knitting machine and had brought it to the shop to show it to the shop owner. The shop owner was telling the inventor that there was no point in inventing a knitting machine and that he should instead invent a sewing machine. His words that a sewing machine was easy to invent and would fetch millions inspired Howe. Besides, it would also ease his wife's work, Howe realised.

Though Howe's family was always on the verge of starvation, he gave up the job and decided to devote himself fully to building the sewing machine. He bought some tools and materials, packed all his belongings and took his wife and children to live at his father's farm. Though his father did not welcome his arrival, he allotted him a loft, where he installed a lathe and began to work on the sewing machine. But luck was not in his favour. A fire destroyed the loft and all his efforts till then came to a nought. Desperate, Howe was wondering as to what he should do next when he ran into George Fisher, an old friend. Fisher was a coal and wood dealer who had then inherited some property. Howe's idea of sewing machine fascinated him so much so that he decided to give it a try. He offered Howe and his family boarding and lodging till his machine was not built and manufactured. In return, he wanted to share half the profit that the manufacture of the machine on a large scale would entail. Howe was more than happy. In December 1844, Howe shifted his family and tools to Fisher's residence.

Thus began Howe's work to invent the sewing machine in a room on the roof of Fisher's residence, free of worry about his family and finance. His previous efforts had shown him that there was no sense in copying the hands of a seamstress as she sew a cloth. He had analysed that his machine should have the following six basic elements: A surface to support the cloth. A feeding mechanism that would move the cloth under the stitching needle as one stitch would follow another. A needle to carry the thread through the cloth. A device which could form a stitch on the cloth. Some mechanism which coordinates all these devices and mechanisms so that all of them perform their functions in proper sequence. In the process of inventing the machine, Howe had hit upon the idea of "eye-pointed" needle, a needle with an eye at the pointed end. Such a needle could go in and out of the cloth being sewn taking the thread along with it. For making a stitch the idea of using two threads instead of one had also struck him, but he could not visualise how a stitch could be made. It was only when he was roaming in the streets one day after having failed repeatedly to devise a means for making the stitch using two threads that he saw a weaver working on his loom and got the crucial idea. He saw that the shuttle in a weaving loom moved to and fro producing a knot in the process. He therefore added a boat - shaped shuttle containing the bobbin of the second thread in his machine.

When the eye-pointed needle, moving through the cloth horizontally, reached the other side, it threw a loop of thread. The shuttle moved through this loop carrying the second thread along, and the needle moved back through the cloth taking the thread along, and thus tightening up the loop. The shuttle also moved back making the lock stitch. The entire process of making the stitch was repeated as the cloth being sewn was thrust ahead by means of a metallic strip pinned to the cloth. In fact, this was the main defect of the machine because after some stitches the cloth was to be unpinned and reset for the next stitches. Nevertheless, it was the sewing machine that worked tolerably well. The size of a modern portable typewriter, it was made mainly of wood, little metal and wires. A wheel and crank arrangement drove the machine on muscle power.

In April 1845, Howe had finished building the sewing machine but it took a few months more to modify it and sew some clothes. The first clothes that he sew were two suits, one for himself and another for Fisher. Fisher was very happy with the performance of the machine as he found the sewing more durable than the cloth of the suit ! The machine had till then cost him 300 dollars, which he felt would soon be recovered. Howe was also thinking on similar lines. His dream of becoming a millionaire was near realisation, he thought. Alas ! He did not know that a few years before him a Frenchman had invented a sewing machine and even manufactured it on small scale but a group of angry tailors, fearing unemployment, had destroyed all the machines and the inventor had died in poverty! In those days, all the garment companies used to employ women for sewing clothes. Clothes were cut on large scale and handed over to these seamstresses who used to do sewing either in the factory or at home at a low price. When Howe showed his machine to tailors, nobody showed any interest in it for the obvious fear that it would replace them. On the other hand, the garment companies found the 300 dollar price of the machine exorbitantly costly as compared to cheap women labour available.

Howe had never expected such a cold welcome to his machine which he had built after so many sacrifices. But Fisher was still confident and felt that nobody cared for the machine because they did not know its worth. And, how was he going to show them its worth ? By actual public demonstration. It was a typically American way of advertising a novel product. A competition was orgainsed at the biggest garment factory at Boston, the Quincy Hall Clothing Manufacture. Five fastest seamstresses of the company were asked to compete with Howe and his machine. Ten seams of cloth of equal length were cut, of which five were handed over to the seamstresses and the rest five to Howe. In the race that followed, heavy bets were made and prizes announced. Howe won the race. He could finish all the five seams seconds before the first seamstress could finish hers! The speed of his sewing machine was 300 stitches per minute.

Although the show was successful the company owner was not convinced that the machine would speed up his production. He was more worried about the heavy initial investment he had to make for the machine. Nevertheless, everybody present in the competition had all the praises for the neat and strong stitches made by the machine. Subsequently, Fisher took Howe to

Washington and obtained a patent for the machine. Fisher had hoped that the patent would secure him some orders. But order for not even a single machine was placed to him. Having spent more than several thousand dollars on the machine without gaining a penny, Fisher lost all hopes. In clear, blunt terms, Fisher told Howe that he would not assist him any more and he should better fend for himself alone. Howe and his family were thus back in the streets.

Helpless, Howe returned to his father's farm and took up a job as a locomotive mechanic to earn for his family. He however did not give up the machine. He went on working on it, making modifications for its better performance. It was during one of his conversations with his father and brother Amasa that the family unanimously agreed that the American garment industry was not accepting the machine because it was basically primitive and that efforts should be made to tap the British industry which was then considered modern. So, Amasa Howe, who was leaving for London by ship, was told to make negotiations with any British manufacturer on behalf of Elias Howe. Unfortunately, he came across William Thomas, a dealer in umbrellas, corsets and leather goods, who agreed to manufacture and sell the sewing machine but cheated him in the bargain. On a small sum he bought the manufacturing rights from Amasa but no sooner he (Amasa) left London he tried to obtain a patent on the machine in his own name ! Thomas could not secure the patent because there were rival claims on such a machine in Britain.

In due course, Thomas also sent a letter to Howe inviting him to London and offering him a job to modify his machine. With no other avenue open to him Howe gladly accepted the offer. He was so sure of making his machine a success in Britain that he even took his family with the aim to settle down there. However, when he reached London he realised that he had made a mistake. Thomas had asked him for his services only to modify the machine so that it could sew corsets. So, after eight months when Howe completed his task Thomas began to mistreat him. Eventually, Howe could not bear it any longer and decided to return home. But, then, he had no money to pay for his passage back to the U.S.A. He therefore mortgaged his patent papers and the original model of his machine and sent his family home. He himself stayed back due to shortage of money. He managed his passage home by serving as a cook on a ship. When he returned home, he found his wife on the death-bed. He could not even afford to reach her on his own money and had to borrow a suit to attend her funeral rites.

Just when Howe realised that his invention was total failure and he would never make a fortune out of it that he saw the first ray of light in the gloom that surrounded him. One early morning he saw a news item in the newspaper which startled him. It was "Sewing machine — the great new invention". Going through the item, he found that his name had not been mentioned any where. When he made further enquiries in the market, he was surprised to find that in his absence from the U.S. his machine had caught the fancy of tailors and garment industry in several cities of the country. Manufactured in half a dozen different models, his modified sewing machine had been advertised in handbills, demonstrated at exhibitions, country fairs and church suppers. There was however no doubt that those modified functioned far more efficiently than his own invention. Howe's blood boiled when he came to know that all the machine manufacturers were making a lot of money. He felt particularly angry with the biggest and successful manufacturer of sewing machines. I.M. Singer Company.

One day Howe personally visited the workshop of the I.M. Singer Company and told the man incharge that the company had been illegally manufacturing his invention. Obviously, the man incharge enquired about his patent rights and other details, which Howe could not give because all those papers were then lying mortgaged in London. In any case he should have those papers, Howe realised if he had even to fight the case in the court of law. Fortunately, he came across Anson Burlingame, who was to visit

London soon. The latter faithfully purchased the patent papers and brought them back to Howe. Had those patent papers not reached him, nobody would have known him today. He would have been forgotten, like so many other inventors of sewing machine.

Howe was then well equipped to fight the case but he had no money to hire a lawyer. Fortunately, at this juncture he met George W. Bliss, a wealthy and intelligent man who was the judge at the sewing competition. Bliss felt that justice should be done to Howe and agreed to support him. Thus began the case in the High Court at Boston. Whereas Howe stressed upon his patent rights claiming that Singer Company had simply modified his original model, the latter dug out old records indicating that sewing machine had been invented by several others and that Howe had simply stolen their ideas ! In other words, Howe had no right over the royalty on the machine being manufactured and sold. Howe however did not lose courage at any stage and fought the case with vengeance. Eventually, in 1854, seven years after he had returned to the U.S.A. that the High Court gave the decision in favour of Howe. The Court also added that Howe had been a benefactor of the mankind due to his invention. The Singer Company had to buy the license from Howe for manufacturing the machine and had to pay 15,000 dollars for the machines already manufactured and sold. Howe was not pacified by this decision alone. He went on to settle the old score with William Thomas. Eventually, Thomas had also to pay five dollar for every machine that he sold in Europe. It was a triumph for justice.

Within a decade of winning his case, Howe became a millionaire, as he had once dreamt. His daily income was 4,000 dollars a day and in 1863 his total income was two million dollars! Along with his brother Amasa and son-in-law, he also started his own Howe Sewing Machine Company at Bridgeport, Connecticut.

In the meanwhile, the Civil War started in the country. Having not forgotten his poverty and exploitation of poor people, Howe voluntarily joined the Northern Army to fight the cause of the negroes, who were being treated as slaves. He however survived the war and died at the age of 48 on October 3, 1867 at Brooklyn, New York. Before he died he had received the "Grand Cross of the Legion of Honour", the highest honour bestowed on a person by the French Government. He also left behind an estate worth two million dollars. For about 20 years after his death his own company continued to produce sewing machines but eventually succumbed to the Singer company. In 1915, Howe was also elected to the Hall of Fame of the Great Americans.

A few months before his death, Howe's patent right on sewing machines was to be renewed. He allowed the patent to lapse so that anybody could manufacture the machine. Indeed, today, sewing machines are still manufactured in millions all over the world. They have today become a household gadget and a best friend of everybody, including tailors.

Enrico Fermi

Enrico Fermi was a great Italian-American theoretical physicist. He was one of the chief architects of the nuclear physics and was awarded the Nobel Prize for physics in 1938 for his outstanding discovery of neutron-induced nuclear reactions.

Fermi was born in Rome on 29th September 1901. He was a brilliant student and obtained his doctorate in physics from the University of Pisa, Italy at the age of 21 with his research on X-rays.

In 1926 Fermi became a full time Professor of theoretical physics at the University of Rome. In 1929, he was elected a member of the Italian Academy. In 1934 he succeeded in making a fundamental discovery in the field of physics. This was the result of 10 years of arduous research. He discovered that when elements are bombarded with slow neutrons, the material becomes radioactive and begins to emit radiation. In this process the element changes into a different material. For example, iron, when made artificially radioactive by neutron bombardment, turned into manganese. In this way Fermi found about 80 new artificial nuclei.

About this time, Italy, under Mussolini, was in the grip of a Fascist government. This affected Fermi directly because his wife was a Jew. So he wanted to leave Italy. Fortunately at this time (in 1938) he was named for Nobel Prize. He was given permission to go to Sweden to receive the award. Fermi left Italy with his family never to return. After receiving the award, he went to U.S. and became a citizen of United States. Subsequently he became Professor for Nuclear Studies at Columbia University.

In 1942, during World War II, Fermi built the first atomic reactor in an empty Squash court in Chicago. Here he set off the first man-made nuclear chain reaction. Later he was instrumental in the development of the atom bomb.

Ernest Rutherford

(1871–1937)

They had been cruising at thirty thousand. The morning sun behind them flung the silhouette of the giant bomber down on a white bed of cloud, a layer of stratus resembling a roll of cotton wool, stretched and spread flat. One moment the shadow would be hundreds of feet below, a midget aeroplane skating like a water beetle over the surface of the cloud, then, as they reached a higher formation, the silhouette would jump up at them, cruise just underneath, with bow and rear cannons clearly visible and the four engine nacelles jutting out in front of the huge wing. There was a break in the clouds; the sea, six miles below, came into view, like the green felt of a billiard table. A second later it had vanished and there was only their own reflection on the cloud. Then, without warning, the cloud ended, nipped off as with a pair of scissors. In front there was nothing but sea and, farther away, the light-brown fringe of coast-line. A few minutes later the sea was behind them and there was nothing beneath but land.

The navigator ordered a turn to starboard and they flew a minute along the new bearing. Then, crackling, over the intercom, the crew heard him shout, "Steady on course—hold it—hold it …" and they strained for a sight of the target. A squat brown huddle of houses came into view. It looked like a village, but that was because it was so far away. Another adjustment and they were coming in on target. It was a quarter past eight, local time,

"She's away!"

The pilot pulled her into a steep climbing turn to port, threw open the four throttles and headed south-west. There was silence on board, absolute silence on the intercom, save only for the drone of the engines.

Suddenly the sky all round them seemed to explode with light. They had instructions not to look back, not to try to see what their strange new, cumbersome weapon was doing to the town behind. Now they knew why. If the light from the explosion were as brilliant as this, ten miles away, six miles up, and facing in the wrong direction, heaven alone would know how bright it was at the point of impact. Heaven and the inhabitants of Hiroshima.

They droned on back to base and it was a fortnight before they learnt what the first atomic bomb had done to Japan. The war was over.

If, in the blinding light, the roar and the million-degree heat of the explosion in the early Monday morning of 6 August, 1945, that caused 60,000 deaths and the almost total destruction of a great seaport, there had been any other sound discernible it might have been the rumble of big Ernest Rutherford turning over in his grave. It had been his most fervent prayer that no one would be able to make practical use of the atomic energy he had discovered and unleashed until the world war — for ever — at peace. The prayer had not been granted.

Ernest Rutherford had been born in Spring Grove (later called Brightwater), New Zealand on 30 August, 1871. His father, James Rutherford, had been brought from Perth, Scotland, as a child by his parents, at a time when New Zealand had a population of only 2,000 white people. James had done his best for that problem : young Ernest had eleven brothers and sisters.

The boy was fond of games and practical jokes and at the same time a voracious reader, but it was some years before he showed the interest in physics and chemistry which was to make him famous. These subjects were hardly taught at the time in small country schools and it is perhaps a stroke of good fortune that while he was impressing his teachers with an all-round ability in Latin, French, history, English literature and mathematics, one of them was subtly influencing the boy in the wonder and excitement of physics, as an out-of-school hobby. The master — a classic master — was W.S. Littlejohn, a Scot from Aberdeen who later went on to become the Principal of one of the most famous schools in Australia.

In 1890, when he was nineteen, Ernest won a scholarship from his little school in the town of Nelson to Canterbury College in Christchurch. He was soon a respected member of the football team (and the largest man in it). Here, too, physics was not encouraged and it was not until he was nearly twenty-one that Rutherford was able to start Advanced Physics. Despite this, he graduated a year later with First Class Honours and within a month was doing experiments as a postgraduate in the damp cellar of the students' quarters. Shortly afterward he published his first scientific paper, "The Magnetisation of Iron by High Frequency Discharges", a work which showed clearly his remarkable ability to make accurate observations with primitive equipment.

In 1894, after a short spell as master in a Christchurch High School — he was not a success, had no idea of keeping order and was puzzled that his class was always in a state of roaring pandemonium — he was awarded an "Exhibition Scholarship" to Cambridge. These "Exhibition Scholarships" had been founded to commemorate the great Exhibition of 1851 and it was with mixed feelings that Ernest learnt he had been selected for one: he had, in the intervals of trying to teach schoolboys, fallen in love with a girl. They agreed, in 1895, to a long engagement and young Rutherford set sail for England.

On arrival in Cambridge he was put to work under Professor J. J. Thomson in the new and splendid Cavendish Laboratory. It was an exciting time to start: a month after he put on his white overall for the first time, W. K. Rontgen discovered X-rays—a first breath of air into the closed room which physics seemed to have occupied since the time of Newton. A few months later A. H. Becquerel showed that uranium compounds produced radiation like X-rays, and a year later, Thomson himself proved the existence of the long-suspected electron. Now the theory which men like Thomson and Rutherford had nursed for years, that all matter had a common origin, had been built, as it were, from the same tiny bricks, became a probability. Rutherford began to experiment with the new "wireless waves" and succeeded in detecting them at a range of half a mile, which was at the time a world record. He caused a stir at the age of twenty-five by demonstrating them during the Liverpool meeting of the British Association. Then, because they seemed worthy only of commercial development, he deserted them for more fundamental matters. He turned his attention first to the problem of conducting electricity through gases, and found that by "ionizing" his gas with a stream of the new X-rays he could pass a sizable current through it. Uranium radiation, he discovered, had quite a different effect and eventually he found that it consisted of at least two different types, which he styled alpha and beta. Alpha rays caused heavy ionization of the gas but hardly penetrated a solid, and beta rays penetrated deeply and ionized less thoroughly.

In 1898, when he was twenty-seven, Rutherford accepted the Macdonald Research Professorship at McGill University in Montreal. He had enjoyed his work at the Cavendish, he had already achieved a great deal, but the chance of a Professorship, his friends told him, should not be lost. He thought briefly of his failure as a teacher in the Christchurch High School and booked his passage to Canada.

He fitted easily into the life of the young university, found his teaching pleasant and not too demanding and continued with his experiments. As a Professor, he could afford a wife, and in 1900 he went back to New Zealand, married his Mary, and brought her halfway up the world to settle in Montreal. They were happy in their new home, even though Ernest spent all day and half the night out of it, in his "wonderful laboratory, the best in the world" as he wrote his mother. It had been handsomely endowed by the Scottish millionaire Sir William Macdonald (who had founded the Professorship Rutherford held). Indeed Scotland can take much credit for Rutherford's work: an Aberdonian introduced him to science in New Zealand, another Scot, McGill, provided him with a university to work in, and a third provided him with a job in it and a fine laboratory.

Soon Rutherford's fame spread beyond Canada and he was approached with offers of appointments in London, Edinburgh, Chicago; but he loved Canada, his wife was happy there, and he could wish for no finer place to work than Macdonald's lab. Young men began to travel to Canada to work with him. One of these was Frederick Soddy; with his help Rutherford in 1902 put forward the revolutionary theory that "radio-activity is a phenomenon accompanying the spontaneous transformation of atoms of radio-active elements into a *different kind of matter*". It was not, as had generally been believed, a chemical reaction. He based his theory on the observation that radio-activity was quite unaffected by heat, cold or chemicals and the vastly more important one, which ultimately produced an atomic bomb that "radio-active change is accompanied by an emission of heat of a *quite different order of magnitude* from that accompanying chemical reactions."

(In fact, as was later proved, each atom produced three million times the energy it might have yielded in a chemical reaction like burning.)

The new theory was treated with scepticism, it cut right across the long-held theory that all matter was indestructible. Nevertheless, Rutherford was elected a Fellow of the Royal Society in 1903.

By 1907 he had decided to return to England and work, in the new laboratory at Manchester University. There had been great strides in technology during the nine years he had been in Canada and now much of the equipment at McGill was obsolescent. In Manchester he formed the opinion that his "alpha rays" were in fact particles. With the help of Hans Geiger's new counting machine, he was able to complete a remarkable *tour de force*, proving not only that they were particles of matter, but that they were positively charged atoms of helium and that 136,000 of them were ejected every second from one-thousandth of a gramme of radium. In a startling experiment with a piece of radium sealed into a thick glass tube, he was able to show the helium gas arriving as if by magic on the outside of it.

In 1908, a year after his arrival in Manchester, he was awarded the Nobel Prize for Chemistry. As probably the most famous physicist in the world—he was still under forty—he may have been surprised at receiving the award for Chemistry, but certainly his work held implications for both sciences.

As in Canada, he attracted the best men to him. Several left Canada to follow him and other front rank men came from the United States, Australia, New Zealand and Germany. In 1910 he was able to propound his new Nuclear Theory, that nearly all the mass of an atom is in the nucleus, positively charged. This is balanced by a quantity of electrons, negatively charged, at a relatively long distance away.

War in 1914 found him lecturing in Australia. He hurried back to the Manchester lab to find, as he had feared, all the younger members of his staff gone. He offered himself to the Government and was put in charge of investigation into Underwater Acoustics, where he was instrumental, by devising methods of listening underwater, in controlling the menace of German U-Boats. By 1917 he felt he could safely go back to fundamentals in his Manchester lab. Not long afterwards he was able to prove that alpha particles bombarding the nucleus of a nitrogen atom liberate the nucleus of a hydrogen atom. In his earlier work with radium he had proved that helium is formed by the *natural* breaking down of that element: now he had achieved an *artificial* transmutation of matter, a thing which alchemists and others had been trying to do for a thousand years.

In 1919 he was elected Cavendish Professor of Physics at Cambridge, in succession to his old superior, J. J. Thomson. He and his wife moved again, and for the last time. His work in this last—but long and fruitful—period of his life was largely devoted to the study of "isotopes". He proved that the different elements might exist in stable forms, with the same chemical properties and nuclear charge, but differing in atomic mass. He reasoned also that in addition to negatively charged electrons and a positively charged nucleus in an atom, then there must be, to maintain equilibrium, a particle with no charge, a neutron. This he found. Bombardment of various elements with neutrons gave rise to numbers of other radio-active substances and it was this development and its pointer to the release of huge energy, that started investigation, using the "235", highly radio-active, isotope of uranium, into the possibility of a bomb and, subsequently, into the use of atomically generated heat for power.

In the last years of his life, which were saddened by the tragic death of his daughter and only child, Eileen, he began a campaign to get for the Cavendish the world's best equipment. Nuclear physics, as this branch of research was now known, could no longer be handled with the primitive tools he had once used. Only the best would do, and this Rutherford got, so that the Cavendish Laboratory, under his direction, became the finest in the world. He became the force behind the Cavendish, doing less research himself, guiding the hands of younger men—and then finding time, each evening, to continue his made-up serial story—reading to his bereaved grand-children.

In 1931 he was created a Baron. He decided not to change his name but to commemorate the small town where he had first been to school, and became Lord Rutherford of Nelson. He cabled his mother in New Zealand, "Now Lord Rutherford more your honour than mine Ernest".

Before he died, in Cambridge in 1937, laden with almost every academic honour his country could give him, he had been awarded degrees by no fewer than twenty-five universities in various parts of the world. He remained to the end the big, modest man who wrote superbly well and spoke-in public-so poorly; the man who, as a colleague said of him on his death: "never made an enemy and never lost a friend"; the man with the resonant, frightening, laugh, who could drop asleep suddenly at a dull patch in the conversation and wake up ten minutes later to resume it. Once when a visitor had been invited to dinner at Trinity College, Cambridge, he had been disappointed not to see the great Lord Rutherford but had enjoyed discussing his favourite topic of farming with a man who had a surprising knowledge of it. The two of them talked till late and then, when they had parted and the visitor was being shown out by his host, turned to shake the proffered hand and said: "Thank you for a most pleasant evening; and just *who* was that fascinating Australian farmer?"

"That," said his host, "was Lord Rutherford."

Edmund Halley

(1656–1742)

The Comet Man

The mysterious comets that wander in space have always been a cause of anxiety, puzzlement and wonder to scientists. Superstitious people consider that a comet exerts an evil influence on human affairs. According to them, whenever a comet appears, natural calamities like terrible floods, famine, torrential rains, earthquakes and even volcanic eruptions soon follow.

A comet has a fiery head, as well as a conspicuously long bright tail stretching far across the sky—sometimes even lakhs of miles long.

A strange comet was sighted in 1944 which consisted of not one but six tails and stretched for lakhs of miles together. In 1811, one Great Comet was sighted the diameter of whose head was 1,70,00,000 kms. and the tail-length was 2,210,000 kms.

The famous English astronomer Edmund Halley first discovered the orbits of comets. He was professor of astronomy at the Oxford University, and also a friend of the great scientist, Newton. Newton had discovered the law of gravitation. It is due to the strong gravitational force of the sun that all the planets revolved round it. Edmund Halley also established for the first time that like other planets, comets also revolved round the sun.

Halley also discovered that such comets had appeared in 1531 and 1607. Halley made a number of calculations. Consequently, he concluded that the comet that appeared in 1531 and 1607 reappeared in 1682 as well.

These figures showed a difference of either 75 or 76 indicating that a comet orbits the sun in 75 to 76 years. Thus Halley's prediction proved authentic. But Halley did not survive to see the authenticity of his prediction. He died 16 years before the reappearance of this comet. In his memory this comet was called "Halley's Comet".

Halley outlined the orbits of all the comets which appeared during 1337-1698, and also collected other relevant details. He also made the first magnetic survey of oceans.

In 1928, another astronomer named Pickering discovered 18 such comets which were at a distance of 7,000 million miles from the sun. On this basis he imagined the existence of one more planet outside the orbit of Pluto.

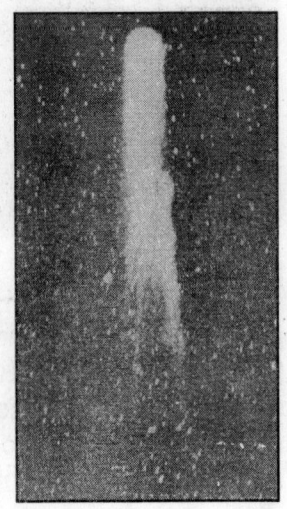

Most of the comets are real wanderers of our solar system and they rarely come out of its edges. Some may move into the solar system from outside also. On entering the solar system they begin to move around the sun. Scientists have so far not decided their source in the universe, as well as their post-disappearing place. The fuzzy head or coma of these comets is composed of tiny rocks, metals, frozen gases and snowballs. Their tail is composed of gases and dust particles embedded in snow. When a comet's tail appears only then it reaches near the sun. The tail follows a comet in the form of a long trail of smoke. When the comet reaches near the sun its tail stretches out due to the sun's heat. And as the comet moves away from the sun, its tail starts shrinking. The most spectacular aspect of a comet is the fluctuating size of its tail.

The oldest picture of Halley's Comet dates back to 684 B.C., which was depicted on Bayeux Tapestry in 1066. Halley's Comet was also probably sighted in 1456 and reappeared in 1759. It was in 1910 that its photograph was taken by a camera.

The Discovering-Campaign of Halley's Comet

Halley's comet appears every 76 years. It is believed that its tail stretches out thousands of kilometres in the space. It is made up mainly of dust and ice particles. When this comet enclosed in its orbit, moves towards the sun it is nearest to our earth and generates a bright light.

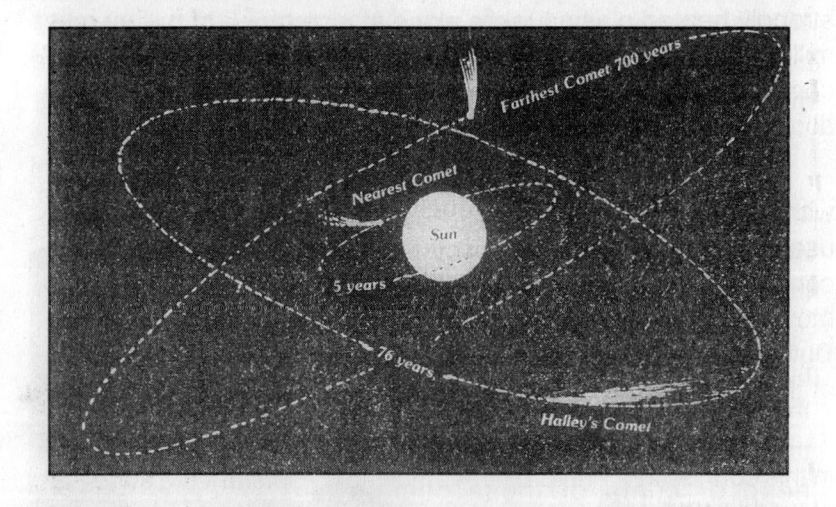

The Halley's Comet was last sighted in March, 1986. A photograph of the comet sent by the Soviet space probe **Veha-2** from a distance of 8,200 kms from its nucleus created quite a sensation among specialists as, contrary to earlier notions, it showed instead of one, two nuclei. The comprehensive, multipurpose programme of the study of the Halley's Comet was fulfilled with the aid of Soviet automatic probes **Veha-1** and **Veha–2,** developed jointly by scientists and specialists of the then USSR, Austria, Bulgaria, Hungary, the GDR, Poland, France, the then FRG and Czechoslovakia.

The nucleus of the comet was seen on TV as a red spot, girdled by an orange ring, then by yellow, green and dark green rings and, lastly by a vast aura—the most rarefied part of the comet's head.

The chemical composition of the gas and dust components of the comet's substance was analysed and electromagnetic field in the comet's neighbourhood and physical processes in its envelope have already been studied.

The subsequent analysis of the observed data showed that the transfer from the undisturbed flows of comparatively cold plasma of the solar wind to the area disturbed by the comet and strongly heated plasma takes place at a distance of nearly one million kilometres from the nucleus. The flow of solar winds completely stops at a distance of nearly 100,000 kms due to some as yet unknown obstacle.

The scientists found it interesting to compare theoretical data with the direct measurement. They even today feel that a lot of useful information can be acquired. Even today they do not completely know about the origin of comets. May be they are moving along very extended orbits or may be, as Jan Oort, the Dutch astrophysicist, believes, they arise even now somewhere at the rim of the solar system.

The **Veha** studies has, in fact, provided enormous information in this connection. The project **Veha** was an outstanding success of world science and an example of international cooperation in space exploration.

Frank Whittle

Cranwell, May 15, 1941: With a roar a new type of airplane took off from the ground. It vanished over the distant horizon like lightning, then climbed up, and came back again in sight. It flew for seven minutes and then going round and round the airfield it descended and made a perfect landing.

A handful of people cheered the successful and eventful flight. "Frank", exclaimed one of them, "it flies." "That was what it was designed to do, wasn't it?" replied curtly Frank Whittle, the inventor of the new type of airplane, now known as jet plane. This curt but suitable remark coming from the inventor is not mentioned here to show how moody he was. On the contrary, how much skepticism about his invention prevailed among the people around him. Moreover, successful test flight of the jet plane was held at the very place where some decades earlier he had predicted its arrival in his essay *Future Developments in Aircraft Design*. At that stage he was ridiculed.

Born at Coventry, England, on June 1, 1907, Frank was the son of an engineer, who was also keen on inventing new things. At an early age he thus became his father's unpaid mechanic. On Sundays he used to be with his father in their small workshop working on lathe and bed. He also became familiar with drawing board, T-square and other *equipments* of a draughtman. In leisure hours he used to read popular science books in a nearby public library. His reading was mostly confined to astronomy, physiology and engineering, particularly aeronautical engineering.

At school Whittle was not considered a scholar. He was, however, brilliant enough to bag scholarships, although he disliked home work. At the age of 16 he joined the Royal Air Force school as an engineering apprentice. Here he built many airplane models and also took part in flying various airplanes. These activities enabled him to get admission in the Royal Air Force College at Cranwell as a flight cadet. He was, however, disqualified at flying airplanes because his logbook noted that he did dangerous flying.

When training at the College came to an end, all students were asked to submit an essay on *Future Developments in Aircraft Design*. Frankly speaking, the aim of the assignment was simply to test the views of the students, and nothing more. In those days the propeller-driven airplanes were quite popular. It was thought that in future they would only be improved upon to increase their speed and efficiency. In their essays, some students therefore wrote that new, more powerful engines to drive airplanes would be invented; others suggested the invention of airplanes with more streamlined bodies and so on. Whittle also wrote the essay but he took the task seriously. He wrote in the essay all that had been cooking in his mind for many years.

In his now historic essay Whittle predicted the invention of what is known as jet-planes. He claimed that faster airplanes travelling at higher altitudes could be built if the then existing piston-driven engine, which drove the propellers of airplanes, could be replaced. He drew, though vainly, the attention of his teachers to petrol-driven turbine which he considered to be an ideal replacement. The essay disappointed his teacher because Whittle has not paid any attention to what he (the teacher) had suggested him to write. He was therefore given poor marks. From that time on his reputation as a crank was also established.

Nevertheless, Whittle secured the second position in final examination and was posted at Wittering as a flight instructor. It was at this place that the seeds of his forthcoming invention were sown. He hit upon an idea which showed him considerable promise in building faster airplanes. Much earlier, in 1791, John Barber had patented a turbine engine but he could not fabricate the turbine because the requisite materials were not available. Also, in 1917, Dr. Harris of Esher had patented the design of a piston-engine for the generation of "jet". Whittle stood on the shoulders of these two inventors but looked far ahead. He combined the two ideas, turbine and jet, to conceive the concept of jet propulsion for airplanes.

Whittle's simple yet fantastic idea was to generate a jet by means of a turbine. A jet of air flowing out of the nozzle of the airplane's turbine would make the airplane move ahead and with equal force, in accordance with the Newton's third law of motion. But, how to generate a jet by means of a turbine? The incoming air to a moving airplane is sucked in through an inlet at the front of the plane. A compressor, which is driven by the turbine, compresses the air. The compressed air is then passed through an ignition chamber where it is heated. The air thus expands and, compressed that it is, forces its way through the exit nozzle at rear of the airplane. In doing so it also drives the turbine which, in turn, drives the compressor. The powerful jet of air produced at the rear moves the airplane ahead.

Whittle disclosed his idea to one of his colleagues, Patrick Johnson, who had studied patent and legal law before joining the Royal Air Force. Johnson felt that Whittle's idea could be moulded into a brilliant invention. He therefore pursuaded Whittle to give it a definite shape on paper, and he himself made efforts to see if the Air Ministry of Britain would be interested in the invention. Unfortunately, the Ministry was reluctant. It claimed the kind of turbine was not possible to fabricate for another decade or so because the requisite materials to bear high stress and temperature were not available.

Although Whittle was disappointed by the Ministry's snappy reply, he, on the persual of Johnson, patented the engine in 1930. Efforts were also made to get some commercial firms interested in the invention. But due to the industrial depression prevalent at that time, even the British Thomson-Houston, the turbine experts, turned down the offer. The company found that it had to risk investing $60,000 to develop the engine. On the other hand, the engine gave Whittle adverse publicity in the Royal Air Force mess, or wherever his colleagues gathered, he became a target for fun. His patented engine was called by all kinds of funny names. Here, again, one flying officer, R. Dudley Williams, was an exception. He had a firm belief that one day Whittle's engine would become a reality. He also helped him in his enterprise. Meanwhile, Whittle's jet-engine patent lapsed. He was hard up for money. He was then a married man and had two growing kids.

Eventually, in May 1935, a firm named "Power Jets Limited" was founded. The founders were Whittle, Johnson and Williams. Subsequently, the British Thomson-Houston also collaborated with the Power Jets to produce a small mail plane. With a handful of technical staff and meagre finance, Whittle started his work in a shed at Rugby grounds, a stone's throw from his house. But, here again, apart from financial worries, another restriction did not allow him to put his whole heart into the invention. The Air Ministry notified him that he could devote only six hours on the engine because he was then studying for engineering at the Cambridge University.

Whittle's first engine was tested on a truck. Although it was not at all perfect, it certainly proved one thing. That it could work. In course of time when the engine began to show its potential, he shifted his workshop to Lutterworth for complete privacy. The Ministry also began to take a keen interest in Whittle's work. After his having secured a first in engineering examination, he was allowed another year for post graduation research and was finally put on the Special Duty List.

In the meantime world War II had overwhelmed the world. Britain was particularly facing a crisis. Seeing the successful test of Whittle's engine in June 1939, the Air Ministry immediately signed a contract with the Power Jets. Gloster Aircraft Company was asked to build the engine as well as the aircraft *Whittle Gloster*. The latter was to be wholly designed by Whittle. This highly ignored invention thereafter became so such precious that it came under the Official Secrets Acts. Meanwhile, wartime research had produced the material required for Whittle's turbine.

In those fateful days when France was overrun and the Royal Air Force had won the Battle of Britain, Whittle was testing his new airplane, unconcerned and patient. In one of the tests his airplane crashed but fortunately no life was lost. The entire project was under such a curtain of secrecy that for several days the police searched in vain the site of the crash for the propeller of the airplane! On occasions, people near the testing site were amazed to see their hats, tea-cups, etc, being sucked up by the overflying airplane. In the beginning all were wonderstruck but it soon dawned upon them that some new kind of airplane was being tested. The characteristic whine of the airplane convinced them that something new was in the air.

Finally, on May 15, 1941, Whittle saw his jet-plane flying in the sky. But he did not stop here. He continued to make finer changes in the engine so that the jet-plane could move faster. He knew that his jet-plane could travel at great altitudes, where air was thin. A propeller-driven airplane cannot travel at higher altitudes because its propellers always need air to push against for moving. The jet-plane thus began to have pressurised cabins and supplies of oxygen for pilot and passengers because at higher altitudes not much air is available for breathing.

In 1943 the first jet-fighter plane *Gloster-Meteor*, a plane faster than *Spitfire*, entered the war scene. In the commercial air travel also Britain's first jetliner *Comet* made its maiden appearance following a successful test flight. Meanwhile Frank Whittle became a household name. He received numerous distinctions, awards and medals from many societies and nations. In 1948 he was knighted and made Air Commodore.

In the end it is rather astonishing to know that Frank Whittle could not enjoy even a single flight in his own designed *Whittle Gloster* jet plane. He had become too valuable to be lost in an accident. It was only when once he got into a standing Meteor without anybody noticing him that he realised how it worked. And when he must have realised that it worked far smoother than any other conventional airplane of the day, all his toil and patience must have been paid off.

Supersonic Travel

A jet plane can attain a velocity greater than the velocity of sound. In fact, jet planes have been manufactured which can attain velocities double the velocity of sound. But at present, these jet planes are not being used for carrying passengers because when they attain such supersonic velocities they produce an explosive sound called "Sonic boom". This boom can shatter windows of houses or eardrums of people residing in the neighbourhood of airports. Military planes are allowed to travel at supersonic velocities only when they are at very high altitudes so that their sonic booms are no more harmful. It is however hoped that in the days to come some gadget to snuff out sonic booms will be invented and installed on supersonic planes. Supersonic travel will then become a reality for every man !

Galileo Galilei

(1564–1642)

Greatest of Early Scientists

Galileo Galilei was one of the greatest of scientific pioneers. His name lives as that of an astronomer who perfected the refracting telescope and showed men the new worlds of the heavens. But his support of the Copernican theory that the earth moved round the sun brought him under the ban of the Church and darkened his life and achievements. His achievements in the studies of mechanics and dynamics were even greater than his astronomical discoveries : he had a happy method of applying mathematical analysis to physical problems and in his work on the laws of motion he paved the way for Newton.

I

I have no wish to go to the moon. I pity the boy who delights in space-helmets and reads so-called 'science fiction' until his eyes pop; these accounts of daring, or demon, scientists have a horrible fascination and are, no doubt, eminently readable. But I deplore nearly all such books and magazine stories for what I think are two very good reasons. This earth upon which we are all placed is surely interesting enough; there are still so many fascinating things to be learnt about it. And all this stuff about 'rockets to the moon', all the stories of flying saucers, men from Mars, space-ships, and so on, are likely to make a boy, who has a genuine interest in things scientific, believe that no exciting discoveries can be made unless he has £500,000 to spend on a rocket or he is given a heaven-sent chance to be sent heavenwards as a cabin-boy or stowaway on some space-ship. And all the atomic work done today, too, adds to the impression that research of any kind is costly beyond the dreams of princes and certainly beyond the means of a schoolboy who has only his pocket money.

What nonsense! I am sure that a boy or girl with an inquiring turn of mind can have more genuine excitement with a not-too-costly microscope or telescope than all space books, all the so-called science fiction, could possibly provide.

The man who was largely responsible for inventing the telescope — an instrument that can wander and probe further into the secrets of space than any spaceship — was an Italian, Galileo Galilei, born at Pisa in Italy in the year 1564.

There was an incident in this great man's youth that I think is particularly encouraging to any young would-be scientist. When he was nineteen and a student in the university of Pisa, Galileo used to go to church in that lovely cathedral with the perilously leaning tower. (It was leaning in Galileo's day, nearly 400 years ago, and it is leaning still.) Now I don't suppose he was the first young man, and I am certain he was not the last, to allow his attention to wander during a church service. Perhaps the preacher was a dull dog. Whatever the reason may have been for Galileo's wandering attention, wander it did.

He found himself watching one of the candelabra hanging down from the high roof of the cathedral, probably upon a lengthy, strong chain. And as he watched it, young Galileo noticed that it was swinging slightly from side to side. Nineteen seems an unlikely age, and the inside of a cathedral seems a most unlikely place, to make a discovery of outstanding importance to all mankind. I don't say that such things can be done by every young man who is not paying all the attention he should to a preacher, but there is no knowing. A quick eye, a lively mind ever asking the question, 'Why?' is much more worth having than all the space-suits and trappings of a dreamy, bogus kind of science.

Galileo watched the candelabrum swinging. He was struck by the fact that the oscillations, no matter what their range, were accomplished in equal times. Ever resourceful (for young men did not carry watches in those days) he checked this observation by feeling that reasonably reliable (clock) there is inside all of us, his pulse. He timed the swings against his pulse. Galileo concluded that the simple pendulum, by means of this equality of oscillation, might be made an invaluable agent in the exact measurement of time. In other words, that is how grandfather clocks were born.

In fact Galileo did not apply this momentous discovery, so important in the accurate, or more accurate, measurement of time, until a good deal later in his long and crowded life. But that morning in the cathedral was one of the important moments in the world's history. All great inventions seem so simple — after the first genius has thought of them.

So much depends upon the careful measurement of time: it is a thing we take for granted today as we take, say, another divinely simple but wonderfully useful invention, the wheel.

Galileo's conclusions from his observation of the swinging candelabrum were about as momentous as the flash of genius that prompted some great unknown in the dawn of history, to say to another hairy, skin-clad man, "Look here, cock, I'm sure we'd shift this great lump a whole lot quicker if we put a round tree-trunk underneath and *rolled* it!"

II

If Galileo had made no further contribution to world knowledge than his observations and application of the swinging pendulum, he would have been remarkable and remembered. But the young man had an irrepressible interest in experiments and great ingenuity in mechanical constructions. He became professor of mathematics at the University of Pisa.

At that time science and medicine had taken few steps forward since the days of Aristotle, the Greek philosopher. Far too much value was placed upon what the ancients had said; far too little importance was given to checking their dogmas against what the keen and experimental eye could see. Galileo was among the first men to ask questions, and to question what the 'authorities' laid down with such authority and, often, pig-headedness. "Aristotle said so, did he? Well, he may be right but let us see for ourselves," was the way—the revolutionary way—in which Galileo approached all things.

As you may well imagine, Galileo was not even in the running for the title of the most popular professor in that university. He aroused much hostility among the teachers and much enthusiasm, too, among the students; students have ever delighted in seeing the pompous, the windy, the pig-headed, cast down from their seats and made to look foolish.

Aristotle had laid it down that a heavier object must fall faster than a lighter body in proportion to its weight. It does seem more plausible, I must say, that if you and I, by some grave mischance, were both to fall from a high window I, with my undoubted greater girth, would deservedly travel more rapidly to my doom.

But Galileo wasn't so certain. That leaning tower of the cathedral at Pisa was to hand, an ideal place from which to drop things. He carried out a number of experiments and proved to his satisfaction, and the discomfort of the Aristotelians, that falling objects, great or small, descend with equal velocity.

For the first, and not for the last time, Galileo was to learn that to be right could be dangerous and that some men love other things more than they love the truth and hate with a deep and passionate hatred the man who dares to upset their treasured and long-held illusions.

In 1591, when he was twenty-seven, and not long after his successful experiments from the Leaning Tower, he deemed it prudent, in the face of the enmity and bitterness he had provoked, to give up his university post and move on to the University of Padua, where he was to teach and work for eighteen years from 1592 until 1610.

I do not think there is much doubt that Galileo brought some of his troubles down upon his own head because he found it difficult to rule his tongue when provoked. He was right, but he was horribly right, maddeningly right; what was more, he said so perhaps a little too often and the bitterness of the discomforted Aristotelians was exacerbated by his telling taunts and cutting sarcasm.

III

When he was at the University of Padua, at the height of his powers, a man in the thirties and early forties, Galileo made his most important discovery. He opened up all the vastness of the heavens to the eye of man by his 'discovery' of the telescope.

I hate to use the word 'discovery', and the word 'invention' because no single man really discovers or invents anything unaided, out of the blue. All scientists are dependent one upon the other, and upon those who have worked in similar fields before them. Science only flourishes where there is a free, world-wide exchange of knowledge. One man advances from where others have left off, rarely from the very beginning: scientists are like boys, climbing one upon the back of the other, until the chap at the top can see something over a wall.

When Galileo was in Venice in May 1609 he heard an account of an instrument for enlarging distant objects upon which a Dutchman was working. What he heard was enough to fire his inventive imagination, and, after much thought and work, Galileo made an instrument, the first telescope of any consequence.

With it he was able to scan the night sky, examine the Milky Way, and even look into the disc of the sun itself.

Those warm, cloudless Italian nights when Galileo sat in the observatory, the first man prying into space, must have been nights of unbelievable triumph. Columbus, high in the shrouds of his cockle-shell boat, looked out only over a great new continent. Galileo was seeing all the planets in their splendour, exploring the mysterious moon. But they must also have been nights of great struggle and mental upheaval for what he saw did not accord with what he had been taught and what he himself, until that moment had believed and taught.

He, too, was wrong.

The moon was not, as he had believed, a self-luminous and perfectly smooth sphere. The Milky Way was a tract of countless separate stars. When, by day, he had examined the sun, he saw the moving spots, and by observing them, reached the conclusion that the sun rotated in about twenty-seven days.

These truths he accepted upon the evidence of his own privileged eyes. He taught and published them. Galileo might have enjoyed a more peaceful old age if he had not been so aggressive in his attitude to those who, like himself, believed otherwise but, quite unlike himself, found it harder to abandon older preconceptions.

IV

In February 1616 Galileo promised to obey Pope Paul V's injunction and undertook thenceforward not to 'hold, teach or defend' the newer doctrines, a step it is hard to understand. It produced peace of a kind, however, for until 1632, when he was seventy, Galileo was not involved in any major conflict with the ecclesiastic authorities. Indeed, he seems to have been regarded with high favour by Pope Urban VIII.

But in 1632 trouble came after the publication of a book, in dialogue form, in which the 'System of the World' was discussed. It gave as much or more offence for the way things were stated as it did for the theories it propounded.

The old Galileo was summoned before the Inquisition. After a long, wrangling trial—during which the sun continued to rotate and the moon to reflect the sun's glory and light—the old man was condemned to abjure by oath on his knees the truth of his scientific discoveries, and his prime revelation that the sun and the earth were not fixed but were, as all is, in constant flux and movement.

He knelt. He muttered the words he had been compelled to utter but, after his recantation, we can well imagine him adding quietly : *"Nevertheless it does move !"*

From the court he went for an indefinite term of imprisonment to the narrow dungeons of the Inquisition, an old, ill, defiant man. Happily the story of his life does not close with the closing of a dungeon door and death coming to him where he could see only a bright star or two through prison bars. Manfully he faced that terrible prospect. But Pope Urban commuted his sentence and Galileo was given permission to reside at Siena and finally at Florence.

At the age of seventy-three, however, he entered a darker world than any dungeon and a lonelier one than any prison cell: he became blind and deaf. Yet, for five more years, so brightly did his spirit burn within him, Galileo continued his researches with unflagging ardour, in freedom.

Galileo had an irascible nature and a tongue like a whip; he had, however, an equally quick capacity to forget and forgive and, what is much more important, to make atonement where he felt that it was necessary. He was the pattern for the scientists who were to come after him and explore the human body, the nature of the atom, and the very substances of the stars, because he resolutely refused to accept statements on the authority of others. He was no negative bone-head. He insisted upon the need to look, see, and experiment.

Galileo was the father of modern scientific method. He lived his life with energy and courage, and made an enormous contribution to our understanding of the world around us.

George Stephenson

(1781–1848)

Builder of The Locomotive

I

At ten o'clock one summer morning, more than a century and a half ago, a cannon boomed in Liverpool and for a moment stilled the excited chatter of many thousands. There were people who had gathered in that great city to watch the opening of the miracle of the day — the railway station between Liverpool and Manchester. The Prime Minister, the famous Duke of Wellington, conqueror of Napoleon, had come from London with many lords and ladies to see for himself this *new wonder of the age*.

Bands were playing. There was one outside the station to play *See the conquering hero come,* there was a band inside the station, and one on the first coach of the Prime Minister's train. Lord Wellington's coach was more like an Indian pavilion than a coach, its sides were elaborately decorated and gilt pillars supported a magnificent red canopy.

Eight trains were to make the run to Manchester. Seven trains carrying guests on one track, and the eighth, driven by George Stephenson himself, was to carry the Prime Minister, chosen guests, and railway directors on the other track.

Safety valves hissed and the engineers waited nervously for the signal. Every precaution against accidents had been taken.

The whole length of the double track had been patrolled to make sure that no careless worker had left a pick, spade or barrow on the lines. Even the cross-over points had been removed to make the line doubly safe. So many people had opposed the railway, insisting it would bring disaster, blight to crops, ruination to farmers, and death through boiler explosions to drivers, that no chances were to be taken.

A crashing of cannon, a blare of music from the band on the State train, and George Stephenson allowed steam to enter the engine cylinders. To a pompous chuff-chuffing the 'Northumbrian' moved ahead, and for a few moments silence fell over the great crowd, to be followed by deafening cheers as one after the other the remaining seven trains, carrying over 700 chosen guests, also began to move off.

Through the gas-lit tunnel that Stephenson had cut through the rock on which part of Liverpool stands, the trains sped, slowly gathering speed, until, when they were finally out into the open country, they were moving faster than the fastest stagecoach had ever travelled. It was an exhilarating experience for many, though some of the ladies were frightened. They seemed to be hurtling along at a very dangerous speed — though any slow goods train of today would probably have passed them with ease.

To show what his engines could do, George slowed down the 'Northumbrian' until one after the other the seven trains on the other line passed him, the guests cheering the famous Duke of Wellington as he sat in his coach, bowing and acknowledging their salutes. Once the last train was ahead Stephenson, a smile on his face, opened the 'Northumbrian's' regulator little by little and very soon they caught first one and then the next of the trains ahead, until finally they were leading again. Such speed had never been thought possible before, and when the trains halted to take on water and fuel at Parkside station half-way between Liverpool and Manchester, there were scenes of the wildest enthusiasm. The 17 miles had been covered in 56 minutes, an amazing speed of 18 m.p.h.

II

It was then, in the midst of the rejoicings and congratulations that tragedy, the first real tragedy of the railways, marred the day; but it gave George Stephenson an opportunity of showing just what his new engines and the railways he had fought so hard for, could do.

Many of the guests had got down on to the railway track, while the engineer of the 'Rocket' drove his train up and down to show how easily he could go either backwards or forwards.

The danger of standing between the rails was not apparent to the excited guests, and a number of them were crowding alongside the Duke of Wellington's coach when suddenly there was a rumble and a hissing. At once there was a chorus of warning cries :

"It's another train. Get off the track! Get off the track!"

The men who a moment before had been awaiting their opportunity to shake hands with the Duke of Wellington turned and to their horror saw the tall white funnel and yellow boiler of the 'Rocket' as that engine bore down on them.

There was an immediate scramble for safety, but Mr. Huskisson, Member of Parliament for Liverpool, either did not realize the danger soon enough or was too petrified to move. Too late he tried to get clear as the 'Rocket' thundered on through the station with a great clatter of wheels and the hiss of steam.

In all that distinguished company there was no doctor. George Stephenson came down to see what was wrong.

"He mun be takken to a doctor at once," he said, his North Country voice stilling the clamour. "Pick him up, gentle-like. I'll tak him to Manchester."

He had to repeat himself, for those standing about were too stunned to understand just what George meant. That the injured man could be rushed to Manchester by train had not occurred to them.

The bandsmen in the front vehicle of the State train were hurriedly ordered out, the injured man was laid gently on a seat, and George climbed on to the 'Northumbrian'. He opened the throttle slowly to spare Mr. Huskisson any jerk, but very quickly the throttle was at its widest, and in the following 25 minutes, that little engine established itself as the fastest piece of man-made mechanism the world had ever known. It hauled the unfortunate Huskisson to Eccles, just outside Manchester, at an average speed of 36 miles per hour. The fastest post chaise would have taken more than twice the time the 'Northumbrian' took, and the journey would not have been half so smooth. That desperate dash to try to save a man's life proved far more than any arguments could, what George Stephenson had been saying for some time, that steam trains could move quickly and in safety. The great engineers of the land had scoffed at the idea of steam trains, other than for pulling wagons of coal at coal mines. Any speed of more than three or four miles per hour, the experts insisted, would result in the boilers of the engines bursting, there would be derailments; cattle in the fields would be so frightened that cows would refuse to give milk; while it was said that the smoke given off by such trains would blight all the crops.

III

That Stephenson had not been too successful in his arguments with the experts was due to his lack of education. Not until he was eighteen years of age did he go to school. Even then it was a very poor kind of school, attended at night after his day's work at the coal mine was done, and he paid a penny per lesson.

The son of a very poor man, George Stephenson was born on June 9, 1781 in a one-roomed cottage in the Tyneside village of Wylam. His father earned 12s. a week, and as there were eight of them in that one-roomed house it is easy to understand why there was no money to pay even the *2d* or *3d* per week which attendance at a Dame school cost towards the end of the 18th century.

But by the time he was fifteen years of age his interest in engines, and his study of them, resulted in his being appointed chief fireman at a coal mine at a wage of a shilling a day. At seventeen years of age he became an engineman, watching one of the water-pumps the task of which was to keep the coal-mines free of the water that drained into the mine galleries.

With no knowledge from books, George learned how his pump worked by taking it to pieces in his spare time, and re-assembling it. His understanding of the engine mechanism finally resulted in his getting his great opportunity.

At a nearby colliery, where at great expense a new shaft had been made, one of the modern pumping-engines was being installed. Built by the great engineer, Smeaton, it was supposed to be the very last word in pumping-engines. George went to watch the men who were assembling the new machinery. He stood and took in the details, then to the amazement and indignation of the engineers he calmly announced that this mighty new pumping-engine was defective and would most certainly not work. It was like a new boy at school explaining that the Headmaster's arithmetic was faulty.

IV

The pump was finally finished and tried. It would not work ! The engineers searched for faults, and could find none. In the meantime the water in the new mine shaft rose day after day. The mine manager grew more and more worried. So long as the pumping-engine was idle, so long was it impossible to get coal from the new mine.

The makers of the engine were sent for. They could find no fault with the engine, nor could they make it work. Finally they announced that the engine would have to be dismantled and taken back to the works. It was a bitter blow for the mine manager, and for all the men who were waiting to work at the new colliery.

Somebody remembered that George Stephenson had said *months before* that the engine would not work, and remembered something else: George had said he could make the engine right, and drain the coal pit within a week. He was sent for. George is said to have been on his way to church when Ralph Dodds, manager of the mine, stopped him to ask if he really thought he could make the new pumping-engine work.

In those days a mine manager was a very important man indeed; he could engage a man or discharge him, and there was no one to say a word about it. George was no doubt startled to have a man like Ralph Dodds seek him out in this way, but he recovered his self-possession after a moment and said :

"Yes, sir, I think I could make the pump work."

The manager of High Pit wasted no time. George was to set to work at once, and Dodds concluded with a promise :

"The engineers hereabouts are all beaten, and if you succeed in accomplishing what they cannot do, you may depend on it I will make you a man for life."

George Stephenson must have been cheered and startled by a remark like that. If he made the engine work, he had won a powerful friend. If he failed, then he would be the laughing stock of the engineers and of everyone else. No one would ever believe again that he knew anything about engines.

On the Monday morning George was at the pit early. The engineering experts were there, and far from pleased at the idea that a youngster like Stephenson, a self-taught engineer, with no workshop background, should be given a chance with the engine which had beaten every one of them.

They stood about, sometimes scoffing sometimes frowning as they watched this youngster doing things to the vitals of Smeaton's engine.

George altered the water injector cock, he raised the cistern by ten feet, he insisted on doubling the steam pressure. One of the alterations had been expressly forbidden by the inventor of the engine, the great Newcomen. One writer says there were bits of wire and oddments about the engine which may have given it a Heath Robinson look, but by Wednesday morning George was ready to try out his changes.

Again there was a collection of experts there to watch, and there must have been a rather frightening silence as George moved over to turn on the steam. Some of the men backed away to a safer distance, remembering that there was twice the steam pressure recommended.

The steam was turned on, and for a few moments there was pandemonium as the watchers scattered. The big pumping-engine came to life with a roar and a clatter which shook the very foundations of the engine-room.

Ralph Dodds, the mine manager, gave an anguished yell as the ground continued to tremble and the clattering roar of the pump made the very air vibrate.

"Why she was better as she was. Now she'll knock the engine-house down", he roared.

There were grins and nods from the experts, but young Stephenson was already busy making an adjustment here, another there, and within a few moments the rattle and roar steadied, died down, and soon the big pump was running smoothly. What was more, water was spurting in a great column from the pump pipe.

All that day, throughout the night, and throughout Thursday, the pump continued steadily lifting water from the flooded shaft. By Friday morning the pit which had been flooded for months, was almost dry, and the same afternoon miners went down to the bottom.

George Stephenson had said he would pump the pit dry within a week, and he had made good his prophecy. What it saved the owners of the High Pit no one today knows; it must have been many hundreds of pounds. They rewarded George with a gift of £10, and a holiday. He had never before possessed and probably had never had a holiday. When he came back from his holiday he was put in charge of the engine he made work, and some years later was appointed chief engine-wright of the group of collieries to which the High Pit belonged. Ralph Dodds had kept his promise, and put George's feet on the first rung of the ladder which was to take him to a where he was acknowledged both in this and other coutries as the master mind where steam engines and railways were concerned.

V

Two great things, in addition to his skill as an engineer made him so successful: his courage, and his ability to plan carefully ahead. He was a man of peace, but if there was a fight to be fought, George would fight, and the story of the one fist fight in his life is worth telling. For here again courage was not enough, so he planned his campaign.

As a brakeman at the Black Callerton pit his job was to work the big engine which lowered the cages down into the mine and brought them up again, sometimes filled with tubs of coal, sometimes with a human load of miners. It was a highly skilled job if the cages were to be worked smoothly.

One day a man named Ned Nelson accused George of jerking them badly. Ned was a great hulking brute of a man, feared by many. After swearing at the young brakesman he threatened to knock his head off for his lack of skill.

"You'll get kicked in the shins as you deserve if you go on like this, " George told him, and there were gasps from the men around. Nelson was startled, then thrusting out his unshaven chin he asked mockingly:

"Kicked did you say? Vho by? Who'll do the kicking?"

"I'll do it," George assured him, at which Nelson guffawed scornfully. He was a burly man in the prime of life, and toughened by years at the coal face while George, though big for his age, was still only a youth.

"There's nobbut one way to settle this !" Nelson roared. " Just one way ! " and George nodded agreement. Mining was a rough trade, and the men were rough. Arguments were seldom settled by words, and Nelson's words were a challenge to fight.

A day was fixed, and Ned Nelson went into special training, taking some time off work to get fit. George's sympathisers could not believe he could have been fool enough to agree to fight. The burly Ned Nelson had had many fights, and more than one unlucky opponent had finished up with serious injuries.

A huge crowd assembled in the Dolly Pit field to watch the contest which was to be fought out according to the rough and ready prize-fighting rules of the day. A rope ring was made and the fighters stripped to the waist.

There were many who sympathised with George when they advanced to begin the fight. George was slight by the side of Nelson. The latter was a great barrel-chested man, with corded muscles which made George Stephenson look even more puny.

Nelson rushed in when the signal for the first round was given. His great arms swung like pistons; his bunched fists, hard from years of work underground, were aiming to knock this young brakesman out for the count in a matter of seconds.

Somehow the big fists only thrashed at the air. George swayed this way and that, moving lithely for round after round while Ned Nelson wasted his strength in mad bull-rushes. In the middle of the third round George did not sidestep when Nelson rushed in. Instead he planted the first of a barrage of shrewd blows.

At the end of the seventh round a gasping, tottering Nelson, one eye closed, and his ribs pink from the hammering he had taken held up his hands to show he had had enough.

George's friends who had thought to have carried him away, perhaps badly injured, crowded round to congratulate him. He pushed them away as the beaten Nelson came across, thrusting out a huge hand :

"Will ye shake hands with me, Stephenson'?" he asked, and George sealed the beginning of a long friendship with a firm and hearty grip.

VI

George won his fight with Nelson as he won many another whether with canal owners who did not want to see a railway robbing them of high rates, or with a patch of quicksand which threatened a tunnel, by sitting down and thinking how his foe had to be dealt with. He knew Nelson was big and strong, therefore he had to keep out of his way until the great strength was worn down. He had planned his campaign and he won.

In much the same way he planned the invention of his safety lamp for miners, and by one of those rare coincidences, he produced a safety lamp practically identical to the one invented by Sir Humphry Davy. George tested a safety lamp at Killingworth, where he was engineer, four days before the distinguished scientist Sir Humphry Davy demonstrated to the Royal Society his miner's safety lamp. The two lamps were almost alike, yet neither man knew of the other.

George Stephenson began work at eight years of age earning twopence a day for looking after cattle. At the end of his life he had seen a dream come true, a land criss-crossed by railways, with trains running at speeds which the great engineers and scientists had once declared were not only impossible, but far too dangerous even to attempt. A committee of the House of Commons, supported by experts, turned down a bill for a line from Liverpool to Manchester, asserting that such a line could never be built over the notorious Chat Moss, a great swamp. The same men lived to see the Moss conquered. For that line George promised to attempt to run trains steadily at ten miles an hour. To try to save the life of Huskisson he drove his 'Northumbrian' at 36 miles an hour, and made the impossible, possible.

George Stephenson did not invent the steam engine; he did not invent railways, for they were in use for horse-drawn wagons when he was a boy. What he did was to put a steam engine on wheels, put it on an improved railway, and make swift, safe travel possible in this and many countries throughout the world. On August 18, 1848, he died. The young man who had paid threepence a week for lessons when he was in his teens had become famous and rich.

Georges Brossard

Champion For The Bug World

Amid the mesmerizing hum, buzz and cackle of the rain forest near Xelha in Mexico, words intrude like false notes in a classic melody. A low, raspy voice carries over the emerald canopy. "Enough is enough. It's time to walk," a middle-aged, leather-skinned man in khaki safari gear tells the young boy he carries on his shoulders.

Realizing that the child has begun to cry, he carefully places him on the ground. The boy looks first at the man, then down at his empty butterfly net, tears rolling down his cheeks.

It is December 1987. The man is Georges Brossard, entomologist and museologist. The child, David Marenger, is a terminally ill patient. The unlikely pair are here courtesy of The Children's Wish Foundation.

Six-year-old David has made a wish to catch the tropical blue morpho butterfly—a beautifully iridescent winged insect. The foundation contacted Brossard, a globe-trotting entomologist who would one day create the Montreal Insectarium, because he was probably the only person who could take the child to the heart of this dense jungle to search for the elusive lepidoptera. After four days of carrying the 22-kilo boy on narrow jungle trails in stifling 38°C heat, Brossard has finally lost his composure. With only two days left to find the blue morpho, he feels the boy isn't trying hard enough.

"You aren't really that sick. Why don't you just get up, and we'll go find that butterfly." Brossard is in fact frustrated with his own inability to locate the blue morpho, something he had thought he could do in a few hours. He has become close to the boy, and even though doctors removed a malignant mass from his brain and were unable to stop the leukaemia from spreading, something tells Brossard that David might not be as sick as the tests say he is.

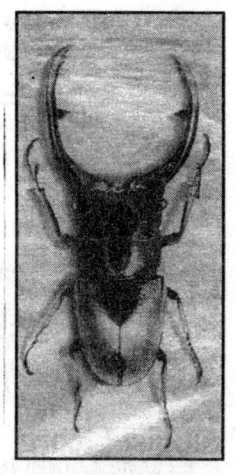

Suddenly, as if to confirm Brossard's thoughts, the boy stands up, glaring defiantly at the entomologist. He picks up his net, puts one foot in front of the other and begins to wobble his way along the path.

Today David Marenger is a fully recovered 19-year-old who remembers fondly the man he once called his second father. And Brosard is still working miracles at 61 — although he insists that it's his six legged friends who make it all possible.

More than 20 years ago, on a one-year sabbatical from his work as a Montreal notary, Brossard was on a beach in Thailand, "thinking about man and the nature of my life. This butterfly flew by my nose, again and again, teasing me. I knew I was supposed to follow it." On the spot, the 39-year-old marched into Bangkok, bought the necessary equipment and followed other insect collectors into the jungle to learn the trade. "I said to myself, ' This is it. I want to devote the rest of my life to insects.'"

Over the past two decades Brossard's passion has taken him to 110 countries to collect as many as 500,000 specimens. But despite his unwavering dedication, few "schooled" entomologists took his collecting seriously. "One day I got angry and decided to go back to school," Brossard remembers. In 1989, he completed his courses in museum organization and administration from the University of Montreal. He also donated his personal collection of 150,000 insects to the City of Montreal.

And one year later, as the proud director-founder, he ushered the public into the Montreal Insectarium. It was the first museum in the world to offer avant-garde, hands-on approach to learning about insects.

Since completing his courses, this Bug Man has become a global authority on establishing insectariums and popularizing entomology. Brussard has helped found seven insectariums around the world, "but I'm hoping to form at least 50 more, he grins.

Brossard has a knack for speaking to crowds and garnering support for his many projects. "Man is crazy, in my mind," he likes to say. "He has honoured all other classes of animals. He's built zoos *for* mammals, aquariums for fish, aviaries for birds. But insects...zero."

Beneath the showman, Brossard is a dedicated scientist sensitive to nature's balance—he collects insects only at the end of their breeding seasons, when they are ready to die. "My collecting never puts populations in danger. I take nothing compared to the number someone kills driving to work every day," he says.

Known for being systematic and organized, Brossard had earned a law degree in 1965. Stephane Letirant, head of collections at the Insectarium, maintains that Brossard's two professions are more alike than most people imagine. "A good entomologist needs a strong memory for order, and so does a successful lawyer," he says.

"Now I've become a lawyer for just one client—insects," Brossard says with a smile. "Let me start by defending mosquitoes. We all know that only the females sting. So already, half of my clients are not guilty. And in defence of the females, they need the blood to lay their eggs. And mosquitoes feed hundreds of species of birds and aquatic animals. So without mosquitoes, nature wouldn't be what it is." He leans back in his chair, satisfied with his arguments.

Energy like Brossard's cannot go unharnessed for long. So he has recently turned his attention to television. An overwhelming success in its first season, Brossard's brainchild, "Insectia," was renewed for 13 more episodes.

"'Insectia' is the ideal vehicle for my mission of reconciling humans with insects," he says.

For One episode, Brossard and the "Insectia" crew travelled by single-engined plane, then by dugout canoe, to the small Venezuelan community of Puerto Nuevo. The villagers there welcomed them with open arms when they saw that Brossard had set his mind on initiating the local children into entomology.

Brossard's first lesson: He quickly taught the children how to catch and handle the giant rhinoceros beetle *(Megasoma acteon)* and showed them the basics of insect anatomy. He then demonstrated the art of using a butterfly net and digging a hole in a tree stump to find beetles. He even taught them how to catch a tarantula. By the end of filming for the episode, Brossard's young apprentices had learned to classify the insects according to their families.

And the episode easily met the show's mandate, which, according to Brossard, is simple: "For every insect that is misunderstood, I give a reason not to hate it."

Whenever he has an opportunity, Brossard will argue that insects are the most prestigious class on this planet, having lived longer than any other. "Ninety-five percent of all species are insects," he says. "It's time for man to recognize their importance."

Not that Brossard doesn't like humans. He's a dedicated family man who cherishes his time with his wife, Suzanne, and his sons, Georges, Jr, and Guillaume. In fact, those close to him say that he's a humanist. "For as long as I have known him, Georges's positive attitude has helped others to be creative, to be involved, to give 100 percent," says Stéphane Letirant. "He's a passionate man, and he's able to transfer that passion to others".

And Brossard is nothing short of a small miracle to the children he meets. Just ask David Marenger's mother about the day Brossard helped her son get his "last wish." Dusk was quickly cloaking the rain forest as man and boy, weary of what had become an impossible task, sat down in the grass. Suddenly, before either of them knew what was happening, a shimmering blue creature lit on David's hand. Reaching out, he touched its velvet wings and examined the iridescent scales. The blue morpho! His dream had come true.

Three days later David walked unaided through the arrival gate at Montreal's Mirabel airport into the arms of his sobbing, dumbfounded family.

"That was a very emotional moment for me," says Brossard. "I couldn't stop weeping along with them."

After his return from the rain forest, doctors were unable to find any leukaemia in David's system. His mother, Yolande Laberge, has no trouble explaining it. "Georges gave my son a dream," she says. "When your dreams come true, it gives you hope. That hope healed David."

Brossard doesn't even try to explain it. He still has a lot of living to do, several billion insects to defend, and a few more butterflies to follow.

Gregor Mendel

(1822–84)

Father of Genetics

Johann Gregor Mendel, a Moravian monk, is called the father of genetics. He experimented with garden peas and the conclusions he drew from his experiments make the foundation of the modern science of genetics. The data and the conclusions drawn from the experiments were presented in the form of a paper and read by Mendel in 1865 before the Brno Natural History Society. It was published in the proceedings of the Society in 1866.

The paper was entitled *Plant Hybridisation*. He read it to an indifferent audience. Mendel scanned the faces of the audience as he was reading it. He expected them to react with the same excitement he himself felt, but what he saw was boredom; they were listening to him only out of politeness. And when he finished the presentation there was hardly a murmur of appreciation. The audience didn't know that they were witnessing a unique moment in the history of science.

Gregor Mendel was born in Heinzendorf in Moravia on July 22, 1822. Moravia was then a part of the Austrian Empire. It is in the Czech Republic now, and Heinzendorf is now known as Hynice. The name given to him by his parents was Johann; he took the name Gregor only when he became a monk.

His father, Anton Mendel was a poor peasant farmer who had to work three days a week for his landlord in addition to working his own small farm. His mother, Rosine, was a gardener's daughter and Mendel probably inherited his love of horticulture from his gardener grandfather.

Johann had two sisters, Veronica and Theresia. Veronica was older to him and Theresia younger.

Johann had his early education at the village school. After this he was sent to an advanced school on the advice of the parish priest, as he was outstanding in school. He was a brilliant student throughout his career, who had to work through high school and university on account of the family's poverty

His father was injured seriously in an accident after which the family's financial condition became desperate.

Johann managed to complete his studies only because Theresia voluntarily gave up part of her dowry to pay for his last year at the university By the time he completed his education the family's resources were totally exhausted. It was at this time that Johann decided to become a priest.

This wasn't such a drastic step as it may appear to modern eyes. In those days there were but a few choices as vocations and many learned men chose priesthood. What probably attracted Mendel was that as a priest he might hope to become a teacher and would have the spare time to carry out scientific studies. Moreover, he was a good Catholic, and faced no moral problems with the religious aspects of the vocation.

Mendel sought help and advice from his physics professor Friedrich Franz. With his help, Mendel was admitted to the Augustinian order of monks, and he joined the monastery of St. Thomás in Brno as a novice, taking the name of Gregor.

This monastery was a great centre of learning and the enlightened abbot Fr. Cyrill Franz Napp, had turned it into a virtual university. The community of monks included a botanist, an astronomer, a philosopher and a composer, all of whom enjoyed good reputations in their respective fields. Fr. Napp encouraged artistic and scientific interests and so Mendel settled in well at the monastery. He attended a four-year theological course at Brno Theological College and was ordained as a priest a fortnight after his 25th birthday.

Fr. Napp sent him to Vienna for higher scientific training four years after his ordination as a priest. At the reputed university there, Mendel mainly studied experimental physics. He also studied statistics, the atomic theory of chemistry and some biology including plant physiology.

Mendel returned to the monastery in 1854 and was appointed as teacher at the 'Modern Technical High School.' He taught there for 14 years, conducting his experiments on heredity during his spare time. In 1856, he began his intensive study of the way heredity works in peas. All the space he had to work in was a small strip of the monastery garden, 35 metres long and 7 metres wide.

We have a picture of Mendel from his former pupils at the school as recorded by his first biographer.

At the school he usually wore the dress of a schoolmaster of those days in Austria — tall hat, frock coat, and short trousers tucked into top boots. He was a jolly, friendly man, "his blue eyes twinkling in the friendliest manner through his gold-rimmed glasses." He used earthy words to describe the sexual process of reproduction in plants. When the boys broke into titters and giggles at this, he told them how silly it was to be embarrassed about natural things.

Mendel's experiments proved that there is something in a plant that determines the properties of its seeds. That something we now call a *gene,* and the property it determines, a *characteristic.* The overall set of characteristics is the *phenotype.* Mendel showed that each characteristic is passed on separately (we now know, by its gene). For example the inheritance of yellowness is independent of the inheritance of smoothness.

Fr. Napp passed away in 1865 and the monks at the monastery had to elect a new abbot. The choice fell on Mendel. There is no doubt that he wanted the job.

All three of his nephews lived near their uncle's residence in Brno and spent Sunday afternoons with him playing chess. With his help two of them became medical doctors. Unfortunately his third nephew died young.

After becoming the abbot he had little time to devote for science. He had to oversee the estates owned by the monastery. He had also to take long journeys frequently : to Rome to pay his respects to the Pope, to Berlin, to Vienna, to the Alps and to Venice.

During his last few years, Mendel was in poor health. He had put on too much weight, smoked small cigars in large numbers, and suffered from kidney ailments. He died in his sleep, at 2 a.m. on the night of January 6, 1884.

H.J. Bhabha, Dr.

(1909–1966)

Father of India's Atomic Energy Programme

The eminent scientist who ushered India into the atomic age was Dr. Homi Jehangir Bhabha. Dr. Bhabha was appointed the first Chairman of the Atomic Energy Commission, set up in 1948. It was also largely due to his efforts that the nation's first atomic research centre, now named Bhabha Atomic Research Centre, was established at Trombay, near Mumbai. Under his expert guidance the nation's first atomic reactor *Apsara* was also commissioned in 1956. In 1945, he founded the Tata Institute of Fundamental Research in Mumbai. From 1945 to 1961, he worked for the development of the atomic energy programme of the country. He is appropriately called the "Father of Indian Nuclear Science".

He was born in Mumbai in a rich Parsi family. His father was a lawyer and his grandfather, the Inspector-General of education in Mysore. His mother was related to the Tatas. Homi grew up in an atmosphere of culture as well as affluence.

Dr. Bhabha had a highly distinguished career and was an exceptionally bright student. He passed all his examinations from the Cambridge university in the highest grades. He took his engineering degree from this university in 1930 and Ph.D. in 1934.

Dr. Bhabha as a scientist of very high calibre, received many national and international awards and honours. In 1942, he was awarded the 'Adams' award by the Cambridge University. In 1948, he received the 'Hopkins' award, and when he was only 32, he was elected a Fellow of the Royal Society, London. In 1951, Bhabha was elected President of the Indian Science Congress. In 1955, he was elected the Chairman of the U. N. sponsored International Conference on the peaceful uses of atomic energy held in Geneva. At this conference, Dr. Bhabha made strong plea for a ban on the proliferation of nuclear power and demanded that manufacture of nuclear devices be declared illegal. The government of India also honoured him with *Padma Bhushan* in 1954. Many awards instituted by several leading science establishments in India after his name. He met a sudden death in an air crash on Mont Blanc on 24 January, 1966 at the young age of 57, while on his way to Vienna to attend a meeting of the International Atomic Energy Agency.

He was a modern man in every sense of the term and wanted India to be at the forefront of atomic energy's creative uses. He used all his extraordinary intelligence for the betterment of India's masses.

Hargobind Khorana, Dr.

Brilliant Biochemist

Dr. Hargobind Khorana is one of the renowned biochemists of the world. He developed a method for the synthesis of deoxyribonucleic acid (DNA) and ribonucleic acid (RNA). For his independent contributions, he was awarded the 1968 Nobel Prize in physiology and medicine, along with M. W. Norenberg and R.W. Holley.

Dr. Hargobind Khorana was born on 9th January, 1922 at Rajpur in Punjab (now in Pakistan). He studied in a village school and distinguished himself right from the beginning by wining scholarships.

He passed his B.Sc. examination from D.A.V College, Lahore and obtained his M.Sc. degree in chemistry in 1945 from Punjab University, Lahore. His main interest was biochemistry He went to Manchester University, in England for higher studies There he worked under Prof. A. Robertson and got his Ph.D in 1948. In the same year he came back to India, but could not get a suitable job. He remained without a job for several months, and finally a dejected, disappointed man, he went back to England for further research. There he worked with Nobel laureate, Sir Alexander Todd at Cambridge University and in 1952 he went to Canada and got married to the daughter of a Swiss M.P.

In 1953, Dr. Khorana was elected as the head of Organic Chemistry Group of Commonwealth Research Organization. He remained in this position upto 1960. In 1960 he went to the United States of America and started working with Norenberg on the creation of artificial life. In the Institute for Enzyme Research at the University of Wisconsin, he developed methods to synthesize RNA and DNA. Due to his research it has now become possible to treat many troublesome hereditary diseases.

In 1970, he joined the Massachussets Institute of Technology (M.I.T.) as Professor of Biology. In addition to the 1968 Noble Prize, he had been honoured with many prestigious international awards.

He was awarded the Padma Bhushan by the Government of India and was conferred with the honorary degree of D.Sc. by Punjab University, Chandigarh.

Henry Ford

(1863–1947)

Great Visionary

The name of Henry Ford raises two controversies of our times—much exploited by school and university debating societies: Is the internal combustion engine a blessing or a curse?

Does mass production make men slaves? Whatever the individual's answer to these questions, Ford is assured of a permanent place in any history of twentieth-century civilization as the first American to realize the potential of the petrol-driven engine and, perhaps even more important, to appreciate that the motor-car would become almost as important to mankind as feet.

As a result of his vision this brilliant engineer built up an industrial empire on which the sun never sets and, in doing so, posed questions of labour and road conditions which, in 2003, we are still struggling desperately to solve.

He was born in 1863, the year of Gettysburg, of farming stock in Greenfield Township, near Dearborn, on the Detroit River which winds through Michigan and divides the United States from Canada. On the day of his birth, 30 July, the fate of the Civil War was still in the balance and the stage was set for the March through Georgia. The telephone, the typewriter, the electric dynamo and, of course, the motor-car were still unknown.

When he was nearly eight Henry went to school in the Scotch Settlement and sat next to a boy named Edsel Ruddiman. He named his first-born Edsel after him. He showed an early facility for repairing clocks and watches but at home on the farm he had to take his share of the inevitable chores, chopping wood, milking cows, learning to harness a team of horses. When he was twelve he was ploughing and doing a man-sized job on the farm. He had no education in science—he got his considerable mechanical knowledge from experience. Engines fascinated him and, after experimenting in a boyish way with water-wheels and boilers, he came upon a traction engine chugging along a road near Detroit. That old tractor was destined to change the course of industrial history. While the driver and his mate were resting by the side of the road, young Henry studied the mechanism—it was his first sight of a self-propelled vehicle apart from a railway locomotive. The power unit of the tractor's steam engine was connected to the rear wheels by a chain and there was belt to transmit the power to sawmill.

He asked the driver how fast his engine could run and was told: "Two hundred turns a minute." That reply, Ford was to say much later, put him into the motor-car business.

His mother, who was originally Dutch, died when Henry was thirteen and three years later he decided to leave the farm and seek work in Detroit. The future multi-millionaire's first steady job was repairing watches for a few dollars a week. During this period he read in a magazine, *World of Science*, that a German, Dr Nicolaus Otto, had invented an internal combustion engine, which had just been patented for manufacture in the United States. The description of this invention so excited him that he joined the Dry Dock Engine Company at lower pay to learn every aspect of the machinist's trade.

Now nineteen, he began to join in the social life of his village, Greenfield, and met a farmer's daughter, Clara Bryant, three years younger than himself. Ford was later quoted as saying that half an hour after he met her, he knew she was the one for him. When in April, 1888, they were married, Henry's father gave him a lot of forty acres on which the young couple built a storey-and-a-half house, with a lean-to for Henry's tools. There he experimented with a farm locomotive and a steam road carriage. This decided him that steam was not the best propellant force for a passenger road vehicle.

One day in Detroit he saw an engine for filling lemonade bottles which had been made on the principle of Otto's gas internal combustion engine. "I've been on the wrong track," he told his wife later. "What I'm after is to make an engine that will run by petrol and get it to do the work of a horse."

To do that meant leaving their comfortable home in the country for the chancy life of Detroit, but the couple didn't hesitate—on 25 September, 1891, their hay waggon rolled east, with Henry bound for a job as a steam engineer with the Edison Company of Detroit. In 1893, when he was aged thirty, he visited the World Fair in Chicago and there he studied the exhibit of a small petrol engine to be used for pumping water. He went home and resumed work on his plans to apply a petrol engine to "a horseless carriage".

Within the year he became chief engineer at the Edison plant at a salary of 100 dollars a month and the Fords took a larger house at 58 Bagley Avenue, Detroit, to live nearer his work. At the back of this house was a brick shed for storing fuel, half for the Fords and the other half for their next-door neighbours.

On his side Ford set up his tools and got down to his work on the petrol engine. A length of one-inch gas pipe served him as a cylinder and in it he fitted a piston with rings. This he attached to a crankshaft by a rod and a hand-wheel from an old lathe served as the fly-wheel. A gearing arrangement operated a cam, opening the exhaust valve and timing the spark with a piece of fibre wired through the centre to do duty as a plug. Contact was made with another wire at the end of the piston and when this contact was broken, a spark leaped across, exploding the petrol. He completed this engine within a week, tested it in his kitchen, made sure it would work, then at once got down to the business of making a two-cylinder outfit.

For the next two years he laboured to apply his two-cylinder engine to propel a bicycle, using the drive wheel of the bicycle as a fly-wheel. He could not get it to work as it always proved too powerful for the wheels. At length he abandoned that idea in favour of a more ambitious one—to make his engine drive a four-wheel carriage. He had taken his first step into the automobile business. The world's No. 1 Ford car was built in the little brick shed at the back of his house in Bagley Avenue.

As it began to take shape his next-door neighbour, Mr Felix Julien, decently gave up his half of the shed and would sit there for hours watching "The Thing" emerge. Parts were made from scrap metal, machined on the lathe and worked on with the tools. The cylinders were bored and fitted in sections of a steam engine's exhaust pipe, the crankshaft was forged at the Dry Docks Works where Ford used four 28-in. bicycle wheels and attached a tiller to the front wheels for steering. The seat was a bicycle saddle. He arranged two belts to offer a choice of speed—10 or 20 m.p.h. There was no reverse—and no brakes.

This veritable "Tin Lizzie", as the later mass-produced Fords came to be called, had its first trial in May, 1896—and its maker overlooked the fact that it was too wide to get it out of his doorway. He just couldn't wait to test it on a road, so he grabbed an axe and smashed down the back wall. After a minor mishap, he drove it once round the block—and the first Ford had made its bow in the city that is now the motor-car capital of the world.

Henry Ford carried on with his job at the Edison plant while he set about designing and building his second car, but the time inevitably came when he was confronted with the choice by his bosses: "You can work on your car or you can work for us—but not both."

Meanwhile several prominent Detroit businessmen discussed a project to manufacture Ford's cars and eventually they formed a company offering him a 10,000-dollar advance, enough to pay for the building of ten cars. Their condition was that he resign his job at Edison's and become chief engineer for the new company, using his design for the cars. By August, 1899, he was out of Edison's and launched on his new career. He was determined that the first car to bear his name and hall-mark would be as near perfect as he could make it and he insisted on improving the carburation system. The ten thousand dollars melted away and his backers began to show their impatience. After 86,000 dollars had been spent they decided to part—and the company's name was changed to "Cadillac Automobile Company".

Ford determined to establish himself so completely with the public that he would find other backers who would set up a company more in accord with his ideas. So he built a racing car and entered it against the most daring driver of the day, Alexander Winton. At the Grosse Pointe Blue Ribbon Track in Detroit, he won an event styled "the first big race in the west" over Winton by almost a mile. Now track champion of the U.S., he had no trouble in finding backers to launch, on 23 November, 1901, the Henry Ford Company with 6,000 shares at a par value of ten dollars each.

By this time there were many others manufacturing high-priced vehicles. Again Ford suffered from production delays, again there was friction with the stockholders and again they and Ford parted. Before he made his third try, he wisely consulted Detroit's successful businessmen and from them learned the know-how of running a sound and prosperous company.

Back he went to the drawing-board to design two powerful racing cars. One called "The Arrow" was wrecked in a race, the other named the "999" after a famous train of that day became the greatest racing car up till then. Its first big win persuaded a Scots coal dealer in Detroit, Alexander Malcolmson, to put his money down on Henry Ford's future. He it was who persuaded an important Detroit figure, Albert Strelow, to provide a factory for Ford and subscribe 5,000 dollars cash for a new venture.

Malcolmson launched with enthusiasm into the task of finding investors for Ford's third—and final—try. Eventually in June, 1903, there were twelve stockholders who between them had raised 28,000 dollars in cash to float the company which is to this day one of the wonders of the world. One of them, James Couzens, who was to be secretary and business manager of the new concern, borrowed 100 dollars from his sister, Miss Rosetta V. Couzens, who was later entered on the books as a stockholder with one share. That 100-dollar loan was eventually to bring her 355,000 dollars.

The infant company struck a snag at the outset. Because of a petrol engine patent granted in 1895 to George Selden, of Rochester, all car manufacturers were required to pay royalties to the lessee of the patent—and there were already a score of tough competitors in the field. To protect themselves these manufacturers had formed an association and had arranged with the lessee of the patent, the Electric Vehicle Company, of

Connecticut, to control the industry. Ford refused to knuckle under and was threatened with prosecution hardly a month after he had started in business. His answer was to challenge the validity of the patent. By the time the court's decision upholding the Selden patent was handed down, it was 1910, Ford's profits had topped the million-dollar mark for the first time and his company had put the Model T—perhaps the most remarkable motor-car in the history of the industry—on the popular market.

The verdict could well have put Ford out of business. Instead, absolutely convinced that he was in the right, he immediately appealed to a higher court. The association warned him that it would be "war to the death" if he did not join. His reply:

"It is said that everyone has his price, but I can assure you that, while I am head of the Ford Motor Company there will be no price that would induce me to add my name to the association."

For nearly two years he had been spending around 2,000 dollars a week defending himself against these manufacturers. Now he was served with an injunction ordering him to stop infringing the patent and his dealers were warned not to offer his "unlicensed cars" for sale. Ford went right ahead building Model Ts. Then the U.S. Court of Appeals gave out its verdict completely upholding all Henry Ford's contentions with regard to the patent. It was an outright victory and even his rivals tumbled over one another to congratulate him on his unyielding fight.

The stage was set for the great leap forward—and it came. Look at the figures of production: 1909—10,600 cars; 1910—18,660; 1911—34,520; 1912—78,440; 1913—168,220; 1914—248,300. Profits soared from 3,000,000 dollars in 1909 to 25,000,000 dollars in 1914. Fantastic as these bounding figures seem, by 1925 Ford was producing 10,000 cars every 24 hours. And, by 1927, when he abandoned the Model T, he had made 15,000,000 of them.

In 1914 Ford exploded a twin sensation—he doubled wages and cut hours to eight a day. It was the beginning of a profit-sharing policy which, the following year, he even extended to all buyers of new Fords, and the apparently unlimited expansion of his organization was halted only by the First World War.

Henry Ford opposed America's entry and he even financed a Peace Ship to carry delegates to a conference of neutrals for "continuous mediation" in the hope of bringing the war to an end. This exercise in practical idealism cost him 400,000 dollars and was the subject of scoffing and ridicule all over the U.S.A. He turned to promoting welfare schemes for his employees, including the Henry Ford Trade School for boys who would otherwise have had no chance of a high-school education.

By the time he could be convinced that motorists were ready for variety and comfort, he had lost the leadership of the industry to General Motors. In 1927 he shut down his vast plant to retool for the Model A and again in 1932 to produce the V-8.

As in the First World War—which he opposed as a pacifist—so in the Second World War—which he also opposed until Pearl Harbour was attacked—he turned over his vast production resources to his country. In the First World War his contribution was in the form of Eagle boats and Liberty motor-cars; in 1941 he laid out a huge bomber factory at Willow Run which eventually produced one complete Liberator bomber every hour and by March, 1945, had achieved the fantastic total of 8,000 of these four-engined warplanes. In the midst of this gigantic war effort his only son, Edsel, died, in the spring of 1943—25 years after his father had handed over to him the presidency of the world's greatest one-man industrial organization. Although his eightieth birthday was only two months off, Henry Ford announced that he himself would take over the presidency again.

When his own time at last come—at Dearborn on 7 April, 1947, in his eighty-fourth year—Ford left his mighty empire intact.

He left, too, the largest trust fund in the world, now known as the Ford Foundation, which is dedicated to the welfare of all mankind.

It is not difficult to summarize his industrial philosophy, which was to cut the price of the product, step up the volume of sales, improve the production efficiency and increase output to sell at still lower prices— and so *ad infinitum*.

It is much more difficult to summarize his controversial genius—so outstandingly brilliant in the mechanics of his age and so woefully *naive* in national, international—and trade union— politics. Perhaps it is kinder to give him an uncomplicated seven-word epitaph: *The Man Who Thought With His Hands.*

Ian Wilmut, Dr. Dolly

He stunned the world with the first clone of an adult mammal. Yet all he'd wanted was to help his father.

It had been a long day, filled with a barrage of calls from the world's press. Inside his office at the Roslin Institute, near Edinburgh, Scotland, Ian Wilmut sighed as he picked up the ringing phone. It was the voice of a woman calling from America, clearly in deep distress.

"Please, professor, I desperately need your help. My little girl has just died. She was only two, yet she fought so bravely against leukaemia. Please, please take some of my daughter's cells and clone her. Give me back my child. I'd do anything to have her in my arms again."

Nothing in Wilmut's training had prepared him for this. He was a scientist, not a doctor. "I'm so very, very sorry that you've lost your daughter," he said gently, "But we cannot return her with this new technology. A child cloned from your daughter's cells could never be an exact replacement. She would be a completely new individual, born later to grieving parents. And she'd be different as a result of a new environment. Hold on to your precious memories—and remember each human is unique. We cannot change that and we shouldn't want to." The announcement of the birth of Dolly the sheep—the first mammal ever to be cloned from a single adult cell—stunned the world. Ian Wilmut and his brilliant team had performed the biological equivalent of splitting the atom, turning the key on an entirely new way of creating life. Yet in January 1997, as the world's press descended upon Roslin, it was that moving call, far more than the media frenzy, that remained in Wilmut's mind. It brought home to the shy, bearded embryologist the huge moral implications of his team's incredible breakthrough.

While in theory it was now possible to clone a human being, that was never part of Wilmut's plan.

For him, Dolly's birth held far greater promise for good. And this hope had its roots in a personal tragedy that Ian had seen unfolding in his own family as a boy growing up in England.

Ian was a sandy-haired, contented child, slightly shorter than his class mates. But what he lacked in height he made up for in enthusiasm. Like his father. Only five foot four inches tall, maths teacher Jack Wilmut had special qualities. You could hear a pin drop in his classroom, parents used to say. The children loved and respected him.

On weekends, Ian and his younger sister Mary would wade through streams, giggling as they fished for minnows. Close by, Jack smoked his pipe. One day Ian saw that his father had a shiny new automatic lighter. "I've wanted one for ages but couldn't afford it," Jack explained. "So I gave up my tobacco for six weeks."

Jack needed that determination. As a student he'd been diagnosed with diabetes and as the years passed his health— and sight—began to fail.

When Ian was 14, he walked into the kitchen one evening to find his father valiantly attempting to cut his mother's hair. Jack's face was flushed with frustration and for the first time, Ian saw his highly capable parent at a total loss. He could no longer see well enough to complete a simple task.

Forced to give up the teaching he loved, Jack struggled to find a job to keep his family from hardship. "I want to be computer programmer," he announced. "I've got all the qualifications, but I need to get on a retraining-for-the blind scheme. The snag is the cut-off point's 38 and I'm too old."

Undaunted, he wrote to Britain's prime minister. "I'm asking Harold Wilson if he can get the rules changed," he told Ian matter-of-factly.

Jack got his place and his new job. But still only in his early forties, he was now totally blind. As Ian left home for university, he took with him memory of how a simple chemical failure in the body could devastate a person's life.

Armed with an agricultural science degree, Ian got a research post at the Animal Research Station in Cambridge. He loved the buzz of a working lab, but above all it was the secret of life itself at its earliest biological stage that held his fascination. His work led to the birth of the world's first calf born from an embryo that had been frozen then transferred to a surrogate mother.

In 1973 Ian moved with his wife Vivienne and two young daughters to the village of Roslin. His research focused on the production of healthier strains of livestock: breeding sheep with high-quality wool, or cattle with a high milk yield. In the early 1980s he helped pioneer an important new method of manufacturing drugs, known as "pharming." By inserting a single human gene into a sheep embryo, scientists could create a "transgenic" animal whose milk produced a human protein for use as a therapeutic drug. The aim was to treat two of the world's most devastating genetic disorders: cystic fibrosis and haemophilia.

Key man–Researcher Keith Campbell's input proved critical.

But the process was frustratingly slow. The Roslin team had to wait for the sheep to have offspring to build up the numbers for this lifesaving work, and there was no guarantee that every lamb in the first few generations would inherit the gene. If only they could clone an adult transgenic sheep, the whole process could be speeded up.

Scientists had already succeeded in cloning from the cell of an embryo. These key stem cells have the ability to develop into any of the 200 to 300 specialized tissues of the body—blood, skin, liver—in much the same way that the stem of a plant branches out to produce leaves or flowers.

But science taught that once a stem cell had "committed" to become, say, skin to liver, it could never return to its earlier life-creating state. Ian could not accept this. Somehow, the way forward for cloning lay in making a copy of an adult mammal. In human terms it was the equivalent of creating a living copy of a person from a single skin cell brushed from an arm. It seemed impossible—yet Ian began setting up his team.

Early in 1991, an affable, longhaired young scientist walked into Ian's office and introduced himself as Keith Campbell. An expert on the complexities of the cell cycle, he spoke passionately about his work. Crucially, Campbell's research into the abnormal behaviour of cancer cells led him to believe that healthy adult cells could be reprogrammed. Wilmut had found the key member of his team. Thus began a partnership which was to rival Crick and Watson's, the duo who had unravelled the structure of DNA in 1953.

Astonishingly, the meticulous research that was to change the very perception of what it is to be human took place in a windowless room no bigger than a cupboard.

The scientist took an unfertilized sheep's egg—smaller than a full stop—then, under a microscope, using a pipette finer than human hair, he removed its maternal DNA, replacing it with the DNA of an adult donor cell from the mammary of a sheep. The aim was to turn back the donor cell's clock so that it would revert to an embryo that would grow into a living copy of its owner.

Such high-precision work needed total quiet—most of the time. One morning an embryologist heard a deafening noise coming from the cloning cupboard. "What's that?" he demanded. "Oh, it's the new Spanish researcher," Wilmut said, smiling. "Talented chap, but an extrovert. He can clone only to Latin American music. It helps him concentrate."

Bringing a family feel to Ian Wilmut's close-knit team were husband and wife Bill and Marjorie Ritchie. Bill, a keen mountaineer, spent his time working away in the laboratory, while Marjorie supervised the collection of eggs from sheep.

Whenever he could, Ian visited his parents. He watched in dismay as his father, still lively in mind, deteriorated physically. A wound in Jack's leg became so badly infected it had to be amputated. It was the worst of many blows Ian had seen his father dealt by diabetes, a condition, which he now knew, was caused by the failure of just one type of cell: the islet cells in the pancreas, which make insulin.

"Only connect," E. M. Forster once wrote. "Only connect." Somehow the very miracle of creation must hold the key to righting these dreadful mistakes of nature. But how? Ian could not fathom it. To make matters worse, he and Keith were struggling to make headway in the lab. Bill and Marjorie were in the centre of many heartbreaking failures as hundreds of cloned embryos were transferred unsuccessfully to surrogate mothers.

Then, in March 1995, Keith came into Ian's office, grinning "I think I've cracked it," he said. The key, he suggested, was to make the adult donor cell dormant—a stage similar to hibernation that stem cells go through before they could impose this shutdown state with a simple procedure: depriving the cell to be cloned of its growth factor, the nutrient serum in which it was being cultured. Then he used a 1250-volt electric shock to weld together the empty egg "shell" with the now-dormant donor cell DNA.

Further experiments convinced them it would work. Cells created this way could develop into a new living creature, just like an embryo created from the fusion of a father's sperm and a mother's egg.

The temptation was to shout the news from the rooftops. "They're trying to do this all over the world and now we've got the key," exulted Ian. But he and Keith knew they had to remain silent until they'd patented their breakthrough, or jeopardize years of work. The following months were tense.

But in July 1996, a perfectly formed, pure-white lamb was born. She was christened Dolly. With no father or mother, her very existence transformed scientific thinking and ushered in a brave new biotechnological age.

Since Dolly's birth, the potential medical benefits of cloning have been acclaimed worldwide. Research into pharmed drugs is now going ahead rapidly, halving the time it will take to reach human trials.

Jonathan Fraser, 18, believes a pharmed drug called AAT, has helped him escape serious lung infections for the past two winters. The youth has cystic fibrosis. He takes AAT, produced by biotech company PPL, to inhibit a substance called elastase which attacks his lungs. Early trials on AAT are still inconclusive, yet Jonathan feels he can now stay alive and fulfil his ambition to become a lawyer.

Cloning technology could also help thousands of people now dying while they wait for a heart or kidney transplant. Within the next few years, research into cloning genetically modified animals could lead to a ready supply of donor organs, enough to meet the needs of thousands of patients, and put an end to the cruel transplant lottery.

But the most exciting research, the medicine of the future, involves therapeutic cloning: culturing and reprogramming a patient's own cells to make healthy replacements for damaged tissue. From six-day-old embryos a unique cell type could be grown—one that retains the ability to form all the body's different tissue types. Ultimately, though, scientists aim simply to take a few of a patient's skin cells, multiply them in the lab and convert them directly to the required cell type.

In ten years Ian Wilmut believes it will be possible to create off-the-peg transplant cells to treat many of today's killer diseases. "All of us must know someone who is suffering right now, from heart disease, hepatitis, blindness," he says. Among the first treatments to be tried on patients would be one for Parkinson's disease—followed eventually by one for the diabetes that so badly afflicted Ian's father, who died in 1994.

For some people, research on an embryo, however small and unviable, seems morally wrong. Nonetheless, Britain's most influential ethical and scientific bodies have welcomed research into therapeutic cloning.

"Stem cell research opens up a new medical frontier," says Professor Liam Donaldson, the UK government's chief medical officer. "It offers enormous potential for new treatments and the relief of human suffering."

In 1999 Ian Wlimut became a consultant to Geron Bio-Med, a merger between Roslin's commercial arm Roslin Bio-Med and the giant US biotechnology corporation, Geron. He travels to California regularly, but his home will always be in Scotland, where he walks the hills.

Nowadays, the cupboard where Dolly was created is visited by politicians, scientists and schoolchildren. They are shown round by Bill Ritchie, whose skilful wielding of a micropipette crafted Dolly at the embryonic stage. "I guess you could say that she's our baby," he comments with irony.

Special permission is needed to visit Dolly herself. But when you enter her barn there is no mistaking the world's most famous sheep. She looks up expectantly, ready to pose for the camera, then, seeing you've omitted to bring one, turns her back in disgust. In April, Dolly gave birth to naturally conceived twins, bringing her family to six, all perfectly normal. She remains oblivious to the fervour her own birth caused—and will continue to cause for centuries.

From the beginning, Ian Wilmut has found himself at the centre of a debate about whether it would ever be ethical to clone a human being. He remains steadfast that, however tragic, there is no human loss that could justify this. On lecture tours he counsels infertile couples to adopt if they can, as he and Vivienne did in 1977 when they wanted a son. They gave a home to Dean, a three-year-old boy who had spent his life in care. Now 26, he lives in Edinburgh and is close to his father.

It is the special value of each and every human life that is important to Ian Wilmut. This remarkable man believes passionately that the cloning technology he pioneered will prove *"a real force for good"*, offering a cure to millions whose lives are currently devastated by disease.

Isaac Newton, Sir

(1642–1727)

Great Genius

"Nature And Nature's Laws Lay Hid In The Night;
God Said: 'Let Newton Be' And All Was Light."

Of those who stand in history among the leaders of the men of genius, Isaac Newton is one. Towering amidst the galaxy of the leading benefactors of mankind arises this colossal figure, in glorified silhouette against the horizon of the ages, sending a glow of radiance down the centuries to the toilers of future. But for his inventions and discoveries the world would have remained a much poorer place.

To him we owe our knowledge of the law of gravity — the great force that governs the sun, the moon, the stars and keeps the heavenly bodies in their courses.

Through his findings that enriched the world he rose to great eminence and won fame as enjoyed by very few of the great men. The grateful world did not fail to acknowledge the greatness of this genius and for twenty-five years he was annually elected president of the Royal Society of London.

To think that this wonderful man who was just a farm boy in the early days of his life and had been at a grammar school for only two years, had, at the age of twelve, discovered the binomial theorem and the principles of calculus, staggers imagination!

He opened the windows to the great mysteries of nature that had been waiting to be uneviled since the creation of the universe. The heavenly bodies beckoned to him and he seemed to understand their silent language.

He was born on December 25, the Christmas Day, in 1642, at Woolsthorpe. His mother's name was Hannah, who little knew that the child she had given birth to would belong to the ages.

He spent the earliest days of his life with his grandmother at Woolsthorpe, when his mother took for her second husband, one Barnabas Smith who was rector of North Witham.

In 1654 he was sent to the Grammar School and we have every reason to believe that he was not a very brilliant student and had little interest in his books.

Physically weaker than the other boys of the class, this boy of a quiet disposition who always kept to himself, easily became the target of the high-handedness of the bigger bullies.

Once the biggest bully of his class struck him, for he was an easy prey, but, to the utter amazement of the whole class, it was the big bully that found himself sprawling upon the floor.

This incident gave him self confidence; he came out of his shell and very soon caught up with the rest of his classmates.

Such minor incidents have often started off men of genius along the path of success and glory.

As a child too he was an observer of what lay around him; he was ingenious by nature.

Near his house there was a water-mill, his most favourite haunt. He liked nothing more than watching how the strong current of the river moved the huge mill-wheel which, in turn, moved the grinding stones that ground wheat into snow-white flour.

After a careful examination and keen observation of the working of the mill, this talented boy made a lovely model of the mill to the great surprise of his friends and neighbours.

Even in such toys one could have failed to see the future inventor of many a thing and idea.

But destiny did not let him bask in the sunshine of the protecting care of his affectionate grandmother for long.

When his step-father died in 1656, his mother, now a widow and left alone, needed a helping hand to look after the farm. She wanted her son back home to be with her. He had to leave school and, when he was just fourteen, was burdened with the worries of a farm. In this way his early career at school came to an abrupt end.

But the boy was made of a different stuff and had no liking or taste for a farmer's life. Nature had designed him for other pursuits and gifted him with other kind of talents.

He was not born for ploughing the field or keeping an account of the cattleheads.

He thirsted for something higher and looked at the horizons—more distant. He longed for knowledge that would raise him from the ground to the ethereal skies wherein float the heavenly bodies, tied to their fixed courses, by some great mysterious law. The heavenly bodies had been sending down their radiant calls, through aeons, to someone who would some day unmask the mystery of the face of the Great Law that governed them.

His lack of interest in what his mother expected him to do may have disappointed her, and we can hardly blame her for that, for she could never have looked into her son's mind and read his interests and aspirations. How could she realize that he was not an ordinary boy and was not made for ordinary pursuits?

He felt so lonely, for there was no one to understand him, to sympathise with him, to read his mind and to know what he thirsted for.

Yes, there was one such man, Issac's uncle, William Ayscough, rector of Burton Coggles, Lancashire. He was also a member of the Trinity College.

He understood what the young boy was thirsting for and he, somehow, persuaded his, sister Isaac's mother, to send Issac back to school, to prepare himself for the College. He worked very hard and was at Trinity College in 1661.

After graduation in 1665, he discovered what is known as the 'binomial theorem'.

To him goes the credit for discovering universal gravitation, the force which holds the heavenly bodies together, guides the earth, the moon and all planets in their fixed paths, according to a highly simple and fixed law. It is also the force which causes a stone to fall to the ground. This force acts between all matter-small or big, earthly or heavenly. For the first time the earth and heaven ceased to be opposed to one another; for the first time the Universe became a single unity — something more astonishing than ever before — in which everything formed a single whole, united by an all-pervading acting force, to which is subjected all visible motion—from our nearest surroundings to the solar system... and beyond....!

His notion was that the force of gravity is not confined to terrestrial objects but acts everywhere and extends to heavenly bodies like the moon.

Some people think that the idea was suggested to him by the fall of an apple is pure legend, but Newton himself was the source of this famous story. According to him, while he was sitting in the orchard on a warm summer afternoon, the fall of an apple to the ground set him wondering why the apple always falls towards the Earth. Thus he reasoned that all matter attracts the rest of the matter to it. If this incident is not a legend, his work on gravitation was inspired without any doubt. From this reasoning he came to the conclusion that the force which attracts the apple towards the Earth is the same force which holds the moon in its orbit around the Earth. This discovery was the beginning of the giant task which culminated in the Apollo's Moon mission.

He used his 'calculus' to show mathematically that the moon was held in its orbit by gravity: otherwise it would move in a straight line at a tangent to its orbit.

Describing mathematically the shape of the earth and the motion of the entire universe and using his theory of universal gravitation he stated that everybody attracts every other object with a force that depends on the masses and decreases with the square of their distance apart.

He returned to Cambridge in 1667 and, in the same year, was selected to a Fellowship of Trinity.

During the next few years he worked on optical work. His experience in optics showed that light consists of many colours.

Constructing telescopes and making investigations into the nature of light took most of his time.

With a telescope having a aperture of an inch he was able to see the satellites of Jupiter in 1668. This telescope was 6 inches long.

His reflecting telescope made him very famous and in 1671 he was elected a member of the Royal Society.

About the 'reflecting telescopes' he writes : "Light consists of rays differently refrangible........... Colours are not qualifications of light derived from refractions of natural bodies, as is generally believed, but original and connate properties, which in diverse rays are diverse . . . . . to the same degree of refrangibility ever belongs the same colour and to the same colour belongs the same degree of refrangibility ."

According to his 'emission theory of light', light was neither ether nor its vibrating motions, but something of a different kind produced from lucid bodies. He says that light originates because of emission, by a luminous body, of minute particle travelling at 300,000 km per second in the empty space. Later on, Thomas Young proved that light travels in waves.

His papers on gravity entitled *Mathematical Principles of Natural Philosophy* were published in 1687.

He passed through difficult times during 1692 and 1694 when he fell ill, suffered from insomnia and nervous disorder. As a result his work was interrupted and the rumour travelled around that he had become mentally unsound.

He was elected to Parliament and, with all his parliamentary duties, he had hardly any time left for his scientific pursuits.

In 1705 Knighthood was bestowed upon him by Queen Anne at Cambridge.

He fell seriously ill in 1727 and died on March 20, 1727. He was buried in Westminster Abbey on March 28.

This great scientific genius and mathematician never married, although there are a few stories of his sweetheart of childhood days. Probably he had no time for romance or marriage. His household was conducted by his married niece; he had three male and three female servants and a carriage.

This great man who was not above middle height and had long and wavy hair, was amiable, hospitable and generous.

A true index to his noble heart is the incident involving his pet dog, Diamond. The dog, one day, entered his study and overturned his lighted lamp. As a result of this, all his papers were burnt to ashes. He had been working on them for years. Any other man would have sentenced the dog to instant death, but Newton, on seeing his years' labour reduced to ashes, only said—

"O Diamond, thou little knowest what thou hast done!"

This great genius was so humble that he remarked :

"I do not know what I appear to the world, but to myself I seem to have been only like a boy playing on the sea-shore, and diverting myself in now and then finding a smoother pebble or a prettier shell than ordinary, whilst the great ocean of truth lay all undiscovered before me."

Though he ceased to exist on this planet, probably he is still wandering among the planets, solar system, the stars, and the galaxies, unveiling their secrets and acquainting himself with their mysteries!

Jagdish Chandra Bose, Sir

(1855–1937)

About seventy years ago a great Indian scientist astonished the world by a number of startling discoveries. By one of these, he paved the way for modern wireless telegraphy and radio broadcasting. By another, he proved that plants too are living organisms. His discoveries were so amazing and so much in advance of the times that they seemed more like fairy-tales than the results of scientific enquiry.

This great Indian who proved beyond dispute that plants have consciousness and feelings was Jagdish Chandra Bose. The great scientists of the world were so impressed by his epoch-making discoveries that they conferred on him the highest scientific honours of their countries. In one of Bose's lectures in England, the great physicist and mathematician, Einstein, himself was in the audience. He was so thrilled and excited that he proposed that Bose should be honoured by erecting his statue in the headquarters of the League of Nations at Geneva.

Let us now see why Einstein was *so* impressed and the scientists of the world showered on him the highest honour.

Jagdish Chandra Bose had proved by actual experiments that plants have emotions and that everything created lives and dies. In explaining his discovery, Jagdish Chandra Bose said :

Hitherto, we have regarded trees and plants as not akin to us, because they are the voiceless of the world. But I will show you that they are sensible creatures—in that they really exist and can answer your questions.

Jagdish Chandra Bose was born on November 30, 1858 at Vikrampur in East Bengal. His father, Bhagwan Chandra Bose, was a Sub-Divisional Officer. Even as a boy, Jagdish Chandra showed a keen inclination towards invention and a strong love of nature. His father noticed his aptitudes and carefully nursed them by providing all facilities to his promising son. He also sent him later to England for higher studies in science.

Bose passed the B.A. Examination in 1874 from the Cambridge University in England winning a scholarship in Natural Science. Next year he took his B.Sc. degree from the London University. After completing a brilliant educational career in London, Bose returned to Calcutta. At this time, he showed definite signs of his many-sided genius. When he was 25, Bose was introduced to Lord Ripon, the Viceroy of India, by Prof. Fawcett who was a famous economist. Shortly afterwards, Jagdish Chandra was appointed Professor of physics in the Presidency College, Calcutta.

Bose had always been an ardent student of science. To work hard, without looking for reward, was for him the key to success. But the reward came soon in the shape of a major scientific discovery. With the publication of his theory on the "*Determination of Indices of Electrical Refraction*" in the journal of the Royal Society of London came the first glory of his scientific career. The learned Society was gratified at the important contribution made by him to the advancement of science. The London University honoured the great Indian scientist by conferring on him the degree of Doctor of Science. Western scientists instantly recognised Dr. Bose's genius. Lord Kelvin was so impressed by Bose's researches on electrical waves that he felt lost in joy. His biographer, Prof. Geddes, records that Lord Kelvin "not only broke into the warmest praise, but limped upstairs into the ladies gallery and shook Mrs. Bose by both hands with glowing congratulations on her husband's brilliant work." This happened when Bose addressed the Liverpool British Association in 1896. Next year, Bose was given the unique honour of addressing the Royal Society of England where he spoke of the nature of electrical waves. It may be mentioned here that only the greatest scientists who have done original research work are invited to address this Society.

In the year 1900, Dr. Bose went to Europe to represent India at the Paris Congress of Science. In that international gathering, Dr. Bose read his paper on *"Response of Inorganic and Living Matter"* which won for him universal praise. Swami Vivekananda, who was present at the Congress, felt full of pride at Dr. Bose's achievement, for he had brought honour to his motherland. Bose was now invited to deliver a series of lectures on his wonderful discoveries. In May 1901, Dr. Bose delivered his second lecture before the Royal Society. In this historic discourse he demonstrated clearly and elaborately the identical nature of reactions in plants and animals. Next year, he gave an illustrated address before the great scientists of Vienna. Prof. Molisch of the Imperial University of Vienna solemnly declared that Europe was indebted to India for the original research initiated by Dr. Bose.

Then followed his lecture tour of America. Learned scholars and scientists of New York, Harvard, Columbia and Chicago listened to him with delight and presented him with eloquent addresses.

Back in India, he worked alone for more than 29 years. He invented delicate instruments to demonstrate and prove his theories. When western scientists, marvelling at his complicated machines asked "Where did you get them made?" Dr. Bose replied with real pride, "In India."

In most of his scientific theories, Dr. Rose was much ahead of his times. He succeeded in sending and receiving signals before Marconi did so in 1907. Much before him, a great French scientist was engaged in investigating plant life but without success. It remained for Dr. Bose to make the amazing discovery that plants have hearts. He invented the crescograph — a delicate instrument for measuring the pulse of plants. His achievements did not stop here. Dr. Bose worked on metals and proved that they too react. Inanimate objects like steel and stone are sensitive and subject to tension. He invented the galvanometer to test the fatigue of metals. More wonderful, he showed that, like animals, metals can be "killed" by poison.

In 1915, he was again given the honour of addressing the Royal Society of England. In 1917, he received Knighthood from King George V. This is one of the highest British honours for meritorious service. In the same year, he set up a Research Institute at Calcutta, to which he made a personal contribution of Rs. 5 Lakhs. The rest of the money - about Rs. 15 lakhs - came from the Government and admirers from all parts of the world. Known as the *Basu Vigyan Mandir*, it has in its laboratories delicate instruments made in its own workshops.

Although a great scientist, Dr. Bose lived like a hermit and gave away his large earnings for great causes. In his will, he donated Rs.15 lakhs for educational, social and humanitarian causes. A lakh of rupees was given for a memorial to Sister Nivedita.

Dr. Bose was a mystic, but to this trait was added the cold precision of the man of science. He was a great patriot and an admirer of the Indian heritage. "If I am to take a hundred more births, each time I would like to be born in Hindustan", he wrote in a letter to Rabindranath Tagore, his friend. He had a special love for India's nationalist song *Bande Mataram*. Whenever he heard it sung, he went into a reverential trance. At the instance of Sister Nivedita, he sent a team of artists to renovate the famous cave-temples of Ajanta and Ellora. It was also in Sister Nivedita's company that he visited the places of pilgrimage in India every summer, including Kedarnath and Badrinath. He passed many days in Swami Vivekananda's *ashram* at Mayavati (Himalayas) where most of his books were written and revised.

In 1925, Bose left for Europe to participate in the League of Nations meeting at Geneva. This was his last European tour. Lecturing before eminent scientists of Europe and England, Bose again demonstrated his sensational discoveries with regard to plant life. On seeing one such demonstration, Mr. Bernard Shaw, the famous British playwright, presented him with a special edition of his works bearing the inscription "From the Least to the Greatest Biologist." The French savant, Romain Rolland, sent him his *Jean Christophe* with the note "To the Revealer of a New World." The editor of the *Spectator* of London organised a lunch in honour of Sir J. C. Bose who "has given a permanence to the Indian civilisation such as no other nation has approached". In the gathering, the greatest literary figures of the time, like Galsworthy, Rebecca, Norman Angel, Yeats, Brown and others came to offer their homage to the great scientist who had enriched human thought. Shortly afterwards, he delivered two lectures at the Vienna University. This was attended by the most eminent scientists and medical men of Europe. The audience burst forth in admiration and warmly congratulated him on his unique achievement. His conquest of the Vienna scientists was so complete that the Rector of the Vienna scientists addressed a letter to the Viceroy of India saying that Sir J. C. Bose's researches had opened out a new gate of knowledge of the highest theoretical and practical importance.

After his triumphal tour of Europe, Bose returned to India in September 1928. This was his 70th year. A great movement was afoot in India to celebrate his 70th birthday and eminent poet Rabindranath Tagore was the pioneering spirit of this celebration. Great intellectuals of India and abroad joined the function which was held at the Basu Vigyan Mandir. The solemn function was inaugurated with that famous song which has now become the National Anthem of India :

Jana-gana-mana-adhinayaka jaya he Bharat-bhagya-vidhata

Rabindranath Tagore presented the great scientist with an address in Verse in which he paid glowing tributes and recounted the glories that Jagdish Chandra brought upon himself and the motherland by his wonderful discoveries.

In November 1937, at the age of 79 the great scientist breathed his last. With him disappeared from the Eastern hemisphere the great that had illumined and led the world of science for so long. Decades have passed since then, but Jagdish Chandra is even today regarded as one of the most original scientists.

J.J. Thomson, Sir

(1856–1940)

The Discoverer of Electron

The British physicist, J.J. Thomson, discovered electron in 1897. He also proved that the matter consists of electrons. This discovery came at a time when scientists were faced with the complex question of the internal structure of cathode rays.

Cathode ray was discovered by another British scientist, Sir William Crookes. He created vacuum in a glass tube by emptying out all air from it and then by discharging a strong voltage of electric current discovered the cathode ray. The German scientist William Roentgen later used the same 'vacuum tube' for the discovery of X-Rays.

In those days a heated debate was going on regarding two subjects. Thomson believed that cathode rays were a cluster of electrically charged particles, whereas other scientists opined that there was a great difference between the cathode rays and electric particles, and they were entirely two different objects.

The opposite view also appeared correct because cathode rays while striking against the walls of the glass tube, used to generate a unique shining light. On the other hand, electrons were not visible to the eyes. Thomson, however, proved by his experiments that cathode rays were not rays but a continuous current of electrically charged particles. Earlier also, Thomson had proved that cathode rays could be deviated by any type of magnetic field or electricity. This meant that the cathode rays were a cluster of electron particles.

In addition, Thomson also calculated the mass of the electron as 1/2000th part of a hydrogen atom. He also calculated the speed of the electron and found that it moved at the speed of 2,56,000 kms/sec.

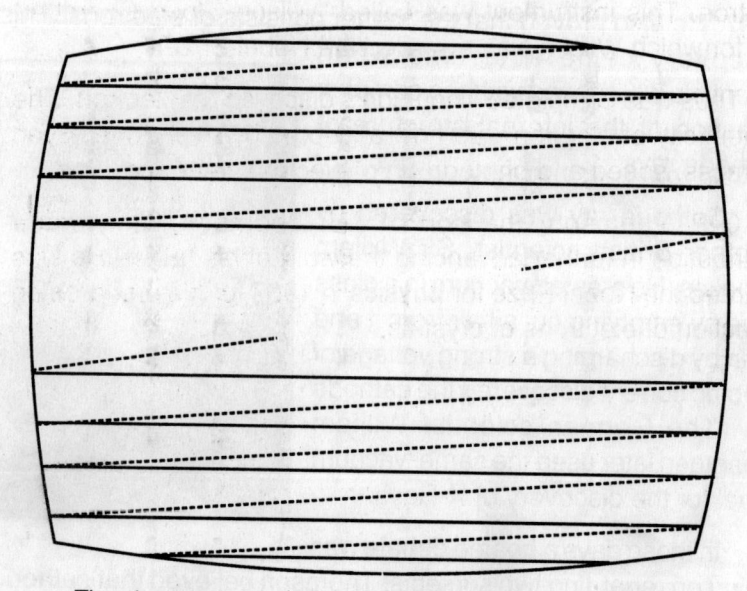

The electron rays are received on the screen of a Black & White TV in zigzag form. Every second 30 frames having 625 lines are formed on it.

The discovery of electron led to the development of a useful device like T.V. The picture tube used in the TV is, in fact, a cathode ray tube only in which electrical particles are deviated at a fast speed. From this deviation we receive the facsimile of the picture which exactly represents the image of man.

However, some scientists were hesitating to accept the rightful place of the electrons. At this stage a student of Thomson, Charles T.R. Wilson, took upon himself the task of taking a photograph of the electron. But Thomson himself thought it impossible to photograph an electron since electron was the 1/2000th part of a hydrogen atom, which is one of the lightest elements.

But after years of persistent hard work Wilson succeeded in inventing an instrument which made it possible to photograph electron. This instrument was called "Wilson Cloud Chamber" and for which Wilson was awarded the Nobel Prize.

Thus was completed Thomson's discovery of electron. The authenticity of electron was now accepted by every one. Even the mass, speed and photograph of electron were accepted.

G.P. Thomson, the son of J.J. Thomson, also made contribution in further advancing the work of his father. He was awarded the Nobel Prize for physics in 1937 for his research on deflection of electrons of crystals.

James Watt

(1736–1819)

Pride of Place Among The Pioneers of Steam

The harnessing of steam power has been one of the scientific romances of the last two hundred years. Men had knowledge of the power of steam nearly two thousand years ago, but it was not until the beginning of the eighteenth century that practical means were devised for its use. Then Captain Savery invented his pump, and Thomas Newcomen followed it up with the atmospheric steam engine. Newcomen was the pioneer, but James Watt was the perfecter, and, in fact, the inventor of the modern steam engine. Watt is the central figure in the romance of steam, which boasts also the names of Travethick, Stephenson, and Fulton, and, later, those of Parsons and de Laval.

It is generally supposed that James Watt was the inventor of the steam engine. He was nothing of the kind. The properties of steam had aroused the curiosity of scientists and inventors for the nine centuries preceding him. Watt was actually an improver who worked out in detail the practical application of the theories of these pioneers of steam power.

In the first century before Christ, Hero, a Greek philosopher living in Alexandria, harnessed steam by means of a curious toy he called the Æoliphile or Ball of Æolus. This machine consisted of a hollow globe of metal, moving on its own axis and communicating with a cauldron of water underneath. The globe was provided with several tubes projecting at right angles to it. The ends of the tubes were closed, but had slits in the slides. When the fire was lit under the cauldron, the steam filled the globe and the impact of the steam against the air coming through the openings in the tubes caused the globe to spin on its own axis.

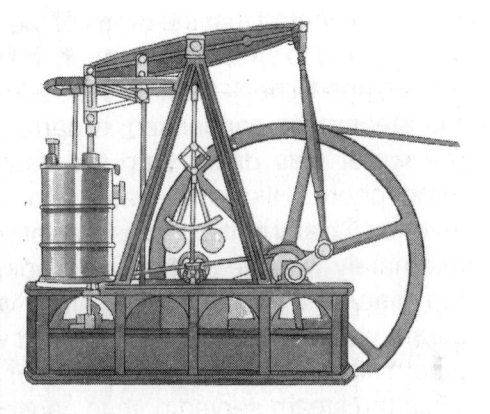

This toy, probably the first machine driven by steam power, aroused in the sixteenth century the wonder of scholars who saw plans of it in Italy. Throughout the next two centuries, speculation on the properties of steam preoccupied scientists throughout Europe. Philosophers would often discuss at their gatherings the properties of the vapours of boiling water. They noted that steam forced within a confined space on to the surface of water would drive the fluid up a pipe to a height which was dependent on the pressure of the atmosphere above. Another peculiarity which aroused curiosity was that steam condensed in a vessel caused a vacuum, thus making it possible for water to be raised from one level to another, owing to the atmospheric pressure, which forced the water to replace the vacuum so produced.

Solomon de Caus applied the first principle to making a steam pressure fountain. His apparatus consisted of a spherical vessel with two pipes; the first to fill the metal globe with water, the other to serve as a vent or jet for the water forced upwards by the pressure of the steam resultant from heating the vessel. Captain Thomas Savery, an English military engineer, contrived to use the second principle for a pump in order to empty the water out of Cornish tin mines. The story goes that Savery, having drunk a flask of Florence wine (Chianti) at a tavern, threw the empty flask into the fire and called for a basin of water in which to wash his hands. He noted that the dregs of the wine in the discarded flask were turned to steam. Thereupon, he took the flask out of the fire and plunged it neck-first into the basin. To his astonishment, after the steam was condensed, a partial vacuum was formed and the water was driven into the flask by the pressure of the atmosphere. His pump was a more elaborate version of the inverted Chianti bottle. It consisted of two cylindrical vessels, filled alternately with steam from an adjoining boiler and water from the mine to be emptied. When either vessel was filled with water, steam was turned on, and the liquid was forced out by another pipe. The process then started allover again in the other cylinder. Thus the steam served both to cause suction, by the vacuum it caused when condensed, and exerted pressure to expel the water. In practice Savery's pump was too slow to be of much service to the Cornish miners, who found that the water came in faster than he could pump it put. Like many of his predecessors he did not realize that the great future of steam power lay in its being harnessed to machinery. The early pioneers thought of steam as something as finite as a primary force.

The first man to employ steam in the sense we understand it was Thomas Newcomen (1663-1729). By the agency of steam he caused certain portions of machinery to move, and applied their motion to work other machines, which in his case were pumps.

Newcomen's engine comprised a vertical cylinder, with a piston working within it. The steam was generated in a separate boiler, from which it was conveyed to the underside of the piston. The piston was attached to a beam, moving on an axis, at whose other end was attached a rod operating the pump. In this way when the piston was depressed, the pump-rod at the other end was raised, and vice versa. The operation of the engine was as follows: the steam, admitted at a little over atmospheric pressure, pushed up the cylinder. The steam, as soon as the piston reached the top of its stroke, was then shut off by hand, and a jet of cold water was turned on to the walls of the cylinder. This last operation caused the steam to condense and a partial vacuum to be formed in the chamber below the piston; atmospheric pressure did the rest, by depressing the piston. Thus, by this slow method, work was done, and the pump was raised and lowered midst much coughing and spurting of steam and water. The engine made twelve strokes per minute and could raise fifty gallons of water from a depth of one hundred and fifty-six feet.

To keep the engine at work, one man was required to attend to the fire and another, usually a boy, to turn alternately two cocks, one to admit steam into the chamber under the cylinder, the other to admit a jet of cold water to condense it. The turning of these cocks proved monotonous work for a boy named Humphrey Potter, who hit on the idea of applying the "up and down" motion of the beam to opening and shutting the cocks. His device, called, picturesquely, the "scoggan," which in north country dialect meant "skulking work," consisted of a catch worked by strings from the beam.

Of Newcomen, who may be said to be the real father of steam power, surprisingly little is known. In those days inventors were frowned upon as schemers, and more the objects of suspicion than of respect. The biographer, therefore, had too little scope to make it worth his while to study closely the background and intimate details of the life of such social benefactors.

Newcomen's early interest in the rise and fall of the lid of steaming kettles has been attributed, as was the case in later years, with James Watt, as the cause of his life-long study of steam-driven machines. Most authorities believe this story to have been invented to add colour to the life of a man whose romance was his achievement. Newcomen was a blacksmith who lived in Dartmouth. At some period he was consulted by Savery, who wanted certain parts of his pump made by a skilled workman. Later, Newcomen took an interest in the work of his employer and it is assumed that he took over the latter's patents at his death. The commercial exploitation of the engine followed under the supervision of a glazier friend of his named Colley. Certain it is that for the next seventy-five years, his "fire engine" was used for keeping the mines free of water.

The work Newcomen did has been too little recognized. Much of the credit for the discovery of the steam engine has fallen to James Watt (1736-1819). In part this claim has been justified, because Watt converted the steam engine from an inefficient machine with but limited uses into a producer of power which could be applied to an infinite number of different purposes. It was used, after he had perfected his design, to pump mines, drive machinery in factories, work flour mills, dig tunnels, build houses, empty ships or mines, and haul masses of goods over ocean, mountain or desert. But at best Watt was no more than an improver of Newcomen's designs.

The exaggerated reputation which has been given to Watt is to no small extent due to a popular story. As a child he showed no promise, and his indolence much exasperated his parents. At the tea table his aunt saw fit to reprove him. "James Watt," said the worthy lady, "I never saw such an idle boy as you are: take a book or employ yourself usefully; for the last hour you have not spoken one word, but taken off the lid of that kettle and put it on again, holding now a cup and now a silver spoon over the steam, watching how it rises from the spout, catching and counting the drops it falls into."

Later commentators, on what is unquestionably an apocryphal story, have tried to prove that while he watched intently that kettle he was working out the theories of thermodynamics. Actually, his interest in steam was probably a pure accident. He became an instrument maker, and by good fortune was allowed to ply his craft within the precincts of Glasgow University. It so happened that a model of Newcomen's engine in the college laboratory was given to him to repair. Although mechanically the engine appeared to be in perfect condition, he found that the engine would not function for more than a few turns. Puzzled, he consulted some of his friends on the properties of steam. One Sunday morning, while he was taking his constitutional after church, he hit on the idea which was to make him the father of the Industrial Revolution. The boiler was too small to operate the engine, therefore the engine was too wasteful in its consumption of steam. The solution was to produce an engine consuming less steam.

Watt found that two necessary conditions for an economical consumption of steam were for the temperature of the condensation chamber to be kept low and the temperature of the cylinder to remain high. He brought this about by removing the condensation process from the cylinder of Newcomen's engine to another vessel, which although it communicated with the cylinder was not a part of it. Steam was exhausted into this neighbouring chamber, which was kept cool with a constant flow of cold water, and there condensed; a partial vacuum was formed by this process without the temperature of the cylinder being lowered. In order to maintain the vacuum and to remove the condensed steam and any air which might have leaked in, Watt added an air pump.

This engine was able to do no more than Newcomen's because it was designed solely as a pumping machine for use in the Cornish tin mines, but it brought about a saving of three-quarters in fuel consumption. Watt's firm claimed as a royalty one-third of the saving effected. The efficiency of the engine was such that at Peacewater, in Cornwall, where the first engine was set up, the mine was emptied in seventeen days, a task which the old engine would have taken months to do.

It was the improved engine Watt produced in later years which made possible the full development of steam power. In order to make the steam engine capable of driving machinery it was necessary for his engine to be converted from a reciprocating into a rotating device. The simplest way of bringing this about was by the use of a crank and flywheel on the same principle as the treadle lathe. Unfortunately, indiscretion on the part of his workmen over his beer, at the Soho ""Coach and Horses," led to his patent being stolen by an eavesdropper.

Balked by theft of his plan, Watt evolved another method of converting the power generated by his engine, by what he called the "sun and planet" motion. This consisted in a cogged wheel attached to the crank connected with the piston. This wheel was enmeshed with another, in its turn fixed to a large wheel which transmitted the power to the machines it drove. Other important refinements of design were the incorporation of the double-acting principle by means of which the power and the speed of the engine were much increased owing to both ends of the cylinder being kept in communication with the boiler; the use of "parallel motion;" an arrangement of links by which the top of the piston-rod is connected to the beam so that it may either pull or push, and is at the same time guided to move in a straight line; finally, he perfected the centrifugal governor, by whose agency the engine is maintained at a uniform speed, and a steam pressure gauge, which is to the engineer what the stethoscope is to the doctor. Thus Watt left the design of the steam engine much as it is today.

In two respects, however, Watt refused to develop his engine by "compound expansion", which would have made it capable of driving a locomotive, and by the use of high-pressure steam. The former, he thought, would endanger the monopoly of his firm; the latter because he was afraid it would imperil public life by causing explosions. His prejudices thus left the way open for Richard Trevithick (1771-1833) and George Stephenson (1781-1848) to develop the steam-propelled locomotive, which brought about rapid transport by land. However, Watt was instrumental in developing steam-driven ships. Robert Fulton, the American designer of the first steamship, ordered his engines from Watt's firm.

Like most inventors, Watt was not fitted for the commercial exploitation of his inventions. He was fortunate in his association with Dr. Roebuck, the master of the Carron iron-works, and

Matthew Boulton, the famous Birmingham manufacturer of silverware. These two men supplied him with the funds he so badly needed for his inventions. At a later date Boulton became Watt's partner, and had it not been for his business acumen it is probable that Watt would soon have given up his experiments. As it was this silversmith, who gave up the manufacture of artistic "Brummagem" ornaments, not only found the vast amount of capital required to finance the installation and upkeep of the engines, but also solved Watt's chief technical problem—he supplied workmen who could manufacture the parts of his engines.

The only available hands in Glasgow, where Watt had made his first engine, were blacksmiths and tinners little able to construct articles outside their ordinary run of business. The men Boulton employed in his works at Soho, just outside Birmingham, were skilled in all manner of metal work, and were thus able to turn out work as accurate as was possible without the aid of machinery. The beautiful self-acting tools made possible by Watt's invention had not yet come into existence, and even with the most expert handwork, carrying out specifications was often an intolerably slow and inaccurate process. Thus no tools were possible without cheap and efficient power, and no general development of power was possible without the exact work machine tools alone could perform. Watt broke the vicious circle.

There is something a little pathetic about the end of Watt. He retired from the vexations of business at the age of sixty-four. Invention now became his besetting passion. The man who had made a vital contribution to human welfare was unable to continue his great work and became a "tinkerer" with "gadgets." His last invention was a machine for copying sculpture! A pointer travelled over the surface of the work and controlled a revolving tool, which cut a corresponding surface on a suitable block of stone. Not long before his death, we find him presenting his copies of masterpieces of the sculptor's chisel to friends as the work "of a young artist just entering his eighty-third year."

He died in 1819 aged eighty-three; his memorial plaque is in Westminster Abbey.

If Savery and Newcomen were the pioneers, it was Watt who invented and perfected the modern steam engine, and he is outstanding among the pioneers of steam-power development. He paved the way for Trevithick, "the father of the locomotive engine," and so for Stephenson, whose railway fame has stolen the thunder from the earlier discoveries. To these men, and to those vaunted heroes such as Branca, Papin, Cugnot, and Murdock, was owed a new age of knowledge and power.

James Young Simpson

(1811–1870)

The Doctor Who Discovered Chloroform

It was James Young Simpson who introduced chloroform in surgery. Less than two hundred years ago, when he was born, surgical operations were extremely painful and sometimes patients even died of pain. James Simpson started the use of chloroform in surgery and saved succeeding generations from incalculable suffering.

He was born on June 7, 1811, at a village called Bathgate in Scotland. His father was a baker. In those days the Scottish countryside was primitive and superstitious. The slaughter of animals was a commonplace event to him; he even witnessed such things as a cow being buried alive because it was believed to be the origin of a disease that had afflicted the village cattle.

Those early impressions, instead of hardening his mind, only served to make it sensitive to suffering and in later life made him wage a battle against pain.

From his earliest days he was good at his studies. He was also practical in everything he did. When he came home from school he would look after the bakery when his mother went out. He was an observant and athletic lad, and was known as the brightest in the village school.

His family had great hopes for him. Poor though they were, they raised enough money to send him to Edinburgh University. He was fourteen at the time. Completing the F.A. Course (Faculty of Arts), he took up medicine in view of his special studies and qualified as a doctor when he was twenty-one. Dr. John Thompson, the professor of pathology, was so impressed by a thesis presented by James for his doctorate that he took him as his assistant. In 1837 Dr. Thompson fell sick and Simpson took his place in the lecture hall. Dr. Thompson had to be on sick leave for a full year which gave Simpson a year's invaluable experience in pathology.

It was at this time that he met Jessey Grindlay, the daughter of a Liverpool merchant and a distant relative whom he was to marry shortly.

The post of professor of Midwifery at Edinburgh University fell vacant in 1839. It was the principal professorship of its kind in Britain, and in view of his special studies on the subject, Simpson applied for the post. He was told that only married men were eligible for the post. Simpson departed immediately for Liverpool, and a month later he returned with Jessie Grindlay as his bride. He applied for the professorship again and was accepted. Thus, he found his life's work.

Now, as a scholar and a professor, Simpson occupied a high position in academic world. But he didn't have the detached outlook of persons holding such positions. He was moved by pain and suffering, especially by the agonies women underwent in giving birth to children.

Some years after he became Professor of Midwifery, news reached England that an American dentist named Morton had used ether as an anaesthetic for extraction of teeth. The need to make surgery painless had long engaged the attention of scientists; Sir Humphrey Davy when he discovered laughing gas, had suggested its use as an anaesthetic earlier in 1800. Morton used ether as it produced a deep sleep. The sleep was not only deep it was long as well. Surgeons wanted the sleep to be long as they were afraid that the patient might wake up in the middle of the operation.

It was at this time that the first-ever operation using an anaesthetic was performed by the famous surgeon, Robert Liston, and the drug used was ether. Simpson was among the doctors. who witnessed it. The medical world was thrilled; Simpson was so impressed that he decided to use ether for midwifery cases too.

Simpson, however found that ether was not altogether a suitable anaesthetic where delivery cases were concerned. He began searching for a more suitable one. Working day and night he stumbled upon the discovery of chloroform in an amusing way.

Working ceaselessly in the day time in the laboratory, he did not relax even after coming home in the evening. He would invite one or two doctor friends to continue the research. The doctors including himself would sit around a table inhaling from their glasses various compounds Simpson had prepared. The compounds were possible anaesthetics and the doctors were guinea pigs for testing them ! This experimenting went on for many days without results. Then one night Simpson suddenly remembered a bottle of chloroform. He had already once experimented with it and rejected it on the ground of being 'slow-acting'. The chloroform was poured into glasses and the three doctors began inhaling it. The effect was amusing. The doctors turned merry and hilarious and everyone in the house joined in the hilarity. The end came suddenly and dramatically. Like in a drinking party the doctors slid down to the floor with a crash one after another. Simpson was the first to recover and to his amusement saw his two friends straddled on the floor in a state of deep unconsciousness.

Dr. Simpson now knew that his search for a good anaesthetic had ended and chloroform was the answer. He immediately began using it in midwifery cases.

Dr. Simpson's views were accepted by the medical profession and the practice of anaesthesia became common in Britain. As a doctor he became so respected that Queen Victoria made him her personal physician in Scotland. Moreover, she herself submitted to anaesthesia when she gave birth to her children.

In the same year the Academy of Medicine in Paris honoured him by electing him as a foreign associate of the Academy, even against the rules of that august body. Other nations followed the example and he was made a member of almost every medical society in Europe and America.

He also made great contributions to gynaecology and obstetrics and his papers on both the subjects are of permanent value. Through his efforts special hospitals were started for the treatment of women in Britain, and other countries followed suit. In this way, he became the saviour of countless mothers.

In surgery also he made his contribution. He introduced a new method of binding arteries. This method is known as *acupressure*.

Sorrows came during his later years. His eldest son, Dr. David James Simpson, died suddenly and a month later he lost his daughter Jessie who was only seventeen. The loss affected him greatly and his health broke down. But he continued to work until he became so weak that it was impossible to work any longer. He passed away on May 6, 1870.

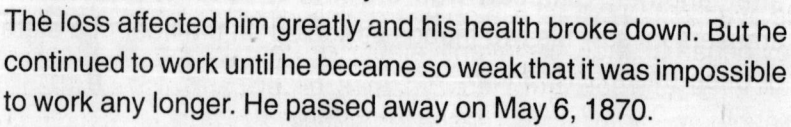

John Baird

(1888–1946)

Great Man of Television

No invention was so easily acclaimed, accepted, and so soon taken for granted as that of television. Its inventor was forgotten even during his life time! That unlucky inventor was John Logie Baird, a little known name today. His is indeed a tragic story of a man constantly struggling against poverty and ill-health, who gained neither fame, honour or money, nor had the satisfaction of seeing his invention in use. The television that is seen today is an electronic modification of the device *"Sending pictures by wireless"* that he invented.

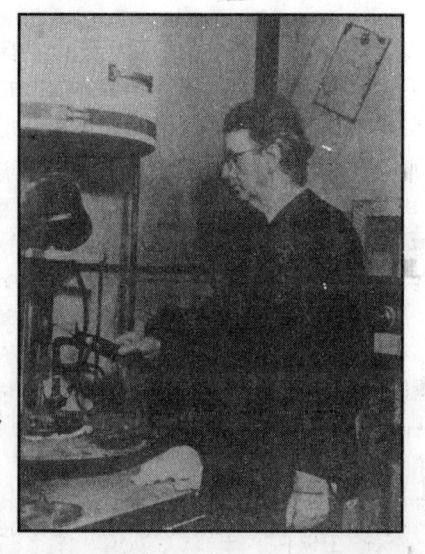

The son of a Presbyterian minister, John was born on August 13, 1888, at Helenburgh, near Glasgow, U.K. From the childhood he had bad health which always worried his parents. He even joined school at a comparatively late age. However, the boy began to show his inventive abilities from the age of 12. He built telephone sets and lines, dynamos, gliders, and also excelled in photography. In fact, his desire to invent a technique of *"Sending pictures by wireless"* can be traced back to his student days at the Royal Technical College, Glasgow. However, he was not allowed to conduct his experiment at the college laboratory. He therefore tried working in the kitchen of his house where he did not achieve any success. But the possibility of inventing such a technique always remained at the back of his mind. When he was 35 and in dire circumstances, the idea gripped him again.

For about ten years Baird had been partly an inventor and partly a businessman, and yet he was failure. He had invented many curious novelties such as under-socks, pneumatic shoes, and had tried selling them without much success. On some occasions he had also sold soaps, fertilizers and jams to keep himself alive. But whenever a slight success was in sight, illness overpowered him and he would soon be back in the same poor state. The decade-old idea of *"sending pictures by wireless"* came to his mind only after he found himself in such a run-down state. He was then looking for an invention that would fetch him a steady flow of money, and in *"sending pictures by wireless"* he sensed such a prospect.

One day when Baird was walking over a cliff, the idea of *"sending pictures by wireless"* struck him. And as he went on considering how it could be invented, he sorted out all the principles involved and the mechanism of the device! It was then left to transform what he had in mind into a working model. When he reached home he immediately wrote a letter to his sister Anne asking her whether he should invent *"sending pictures by wireless"* or he should continue to sell the glass safety razor which he himself had invented. His sister advised him to play safe and continue with the business of safety razor but Baird disagreed. He began the preparations for the experiments that were to bring to the world a novel device.

Although Baird's idea was a novelty, it was not something beyond the imagination of scientists and engineers of the time. Heinrich Hertz had already discovered radio waves and Guglielomo Marconi had invented wireless telegraph to send messages. *"Sending pictures by wireless"* was therefore considered to be the next step, whose invention would be announced any day.

With whatever meagre resources he had, Baird began in 1923 his experiments in his bedroom. For his experiments he gathered from the junk shop the required equipment and even made use of his furniture. It did not take him long to achieve the initial success, which convinced him that *"sending pictures by wireless"* was possible. He was able to transmit the shadow of a Maltese cross across a distance of more than half a metre. For further experiments he however needed money. He therefore gave a demonstration of his experiments to the press in the hope of getting fame and also a patron, who would give him money to continue his experiments. The demonstration, however, did not impress the press. He got some publicity but no patron. Instead, a few days later, he received about 50 pounds as a gift from his father!

Again in March, 1925, Baird gave a public demonstration of his invention for the sake of money. The occasion was the birthday week of the owner of Oxford Street Store, Gordon Selfrige. Pictures of paper masks with slits for eyes and mouth that could open and close were shown transmitted by wireless. It was, in fact, the first public demonstration of TV. However, the public took it as a stunt to get publicity. A few, on the other hand, were apprehensive that a device such as this, which could see though brick walls and around corners, would destroy privacy. Some scientists also watched the show but with considerable scepticism.

Baird soon became tired of the shows and gave them up. He, however, continued to nag the press to give him enough publicity, so that could get some patrons. In due course, he earned the reputation of a harmless crank among the pressmen. His bedraggled appearance, long unkempt hair and an air of absent mindedness were considered good enough only as a cartoonist's scientist. "Show us a living face — and you can have it," the pressmen finally told him.

Eventually, on October 2, 1925. Baird succeeded in transmitting the face of "Bill", an old ventriloquist's dummy, in considerable details. Immediately, he ran downstairs to fetch a living face. In the stairs he met William Taynton, a bearer working in the film company below Baird's room, whom he literally dragged upstairs. Puzzled, Taynton allowed Baird to make him stand before a complicated equipment. But when Baird reached the receiving screen in the next room, he found nothing there. He immediately understood the reason, ran downstairs and caught Taynton slipping away. The bearer told him that he was afraid to stand before the complicated equipment which also threw a glaring light. Baird therefore thrust a few coins into Taynton's hands and asked him to stand where he wanted him to. So Taynton's face became the first to be televised — and also paid for. But, how was it televised?

The working principle of *"sending pictures by wireless"* is simple. Every picture, or for that matter, any object can be considered to be composed of dots. The tone of the picture depends on the density of the dots. For instance, as the density of dots increases, the tone changes from a lighter shade of grey to a darker one, and finally to black. Now, any picture is composed of lines (take horizontal ones), each, in turn, is composed of dots. The sharpness of a picture depends on the number of lines a picture is composed of. The more the lines, the sharper the picture grows. For instance, in modern TVs the screen is composed of more than 500 horizontal lines. Baird's maiden TV screen was however, composed of only 30 lines.

If the dots in picture could be transmitted systematically, one by one, from the left to the right of the picture, the line would be reproduced. Similarly, all the lines, from the top to the bottom of the picture, could be reproduced systematically so as to reproduce the complete picture on the screen. The lines should, however, be so speedily reproduced that out eyes should be able to see them together, simultaneously as a complete picture, and not as separate lines. In modern TV an electron beam in the

Cathode Ray Tube reproduces the entire picture on the screen. Baird used a far cruder mechanical system to reproduce a picture.

Baird used a Nipkow disc in which holes were present in such a manner that in one rotation of the disc the holes scanned a picture completely from top to bottom. The dots, the spots of light, weak or bright, that the disc broke the picture into; were converted into electrical signals, weak or bright, by means of a photo to re-compose the entire picture. This was how Baird's *"sending pictures by wireless"* worked. When he demonstrated his invention before the press in 1926, he became world famous overnight.

However, Baird did not stop here. Now well equipped, he carried on his experiments to extend the range of his invention. In June 1927 he was successful in transmitting to Glasgow pictures of people sitting in a drawing room in London. One year later, he transmitted pictures from London to New York. In 1929 his efforts were crowned with success when the British Broadcasting Corporation adopted Baird's techniques for transmission of pictures from their London station. Subsequently, sound was also added to make it a complete live TV. Baird received wide appreciation when he televised for the first time the horse race at Derby in June 1931.

Meanwhile TV had made a beginning as an entertainer not only in Britain but also in the U.S.A. At the British Broadcasting Corporation there were weekly transmissions of a variety of entertainment features.

When the TV was about to become popular among the masses that its downfall began. The viewers became dissatisfied with the pictures that appeared on their receiver screens. The images were orange in colour because of the neon light, and strained the eyes when viewed for a long time Further, there was flickering of the image all the time because TV produced 25 picture in two seconds which was not enough for the persistence of vision. In short, in comparison to the films that were then becoming popular, Baird's TV stood nowhere. Gradually, the telecasts were put off the air and Baird was forgotten.

It was only when Cathode Ray Tube, the wonder tube of electronics, was invented that TV again made a dramatic comeback. The Cathode Ray Tube made the image steady and sharp. For reasons of his own, Baird, however, never felt that Cathode Ray Tube could replace the mechanical system he had invented. It proved to be a costly mistake for him, and it was too late when he realised it. He died as a non-entity in 1946 after catching a chill. Before he died he had also devised the first satisfactory colour TV and 'noctovision," a device to see in the dark.

Two Way Television

The scene that appears on the TV screen is what is being shown in the studio of a TV station. Change the channel, and the scene will also change to what is being shown in another studio. But imagine that one can not only see what is being shown in a TV Studio but also in any body's home, office, factory, etc. And the other person will also be able to see at the same time what is happening wherever you are ! This *"Two way television"* is likely to enter our homes, offices, factories — in fact, everywhere — very shortly. It will work just like a telephone. One has to dial a number, and when the person at the other end will switch "on", you can see each other and talk face to face. It will reduce a lot of travelling. Without leaving one's home or office, one can talk face to face with any one, whether he is a friend, doctor, grocer, banker, etc.

John Flamsteed

(1646–1719)

Father of Modern Astronomy

"A prophet is not without honour, save in his own country," and truly, few Englishman today know or honour the name of John Flamsteed. Yet he was the father of modern astronomy, the first Astronomer Royal, who corrected all the astronomical tables in use in the seventeenth century, who provided Newton with the observation that made that great man's own discoveries possible. Racked with pain, handicapped by poverty, cramped and thwarted by his contemporaries, even by Newton himself, John Flamsteed gave to the world in his life's work one of the greatest contributions to practical astronomy ever made by one man.

In the summer of 1660, the year of Charles II's restoration, a fourteen-year-old boy went for a bathe, and caught a cold. That cold gave Britain her first Astronomer Royal, Greenwich Observatory, and all that these names stand for. The boy's name was John Flamsteed; and he was a scholar at the Derby Free School. His father, Stephen Flamsteed. was a maltster. The boy had been born at Denby, five miles away, on August 19, 1646, and three years later his mother had died.

The cold brought other ailments in its wake. The chief of them was rheumatic affliction of the joints. Young John Flamsteed became very ill; rheumatism crippled him; he grew unable to walk to school, and left in May, 1662.

From his fourteenth year to his death Flamsteed struggled against illness and physical agony, but the illness gave him opportunity. He could not go to school, so he began to teach himself. He picked up a copy of Sacrobosco's *De Sphaera,* and read it between the agonizing bouts of his illness. The book interested him and attracted him to astronomy. He read other works on the same subject, and at once began to practise. He observed a partial solar eclipse in September, 1662. He made himself a rough quadrant. He compiled a table of the sun's altitudes. Already, the sick, crippled boy was an astronomer. He tells us how he worked away under the discouragement of friends, the want of health, and all other instructors except his better genius."

Stephen Flamsteed tried all kinds of treatments to get his son cured, but there was little help in medicine. In 1664 he sent John to a man named Cromwell, who was "cried up for cures by the nonconformist party." But Cromwell could do no better than the physicians, and in the next year John Flamsteed went to Ireland to be "stroked" by Valentine Greatrakes, who had acquired tremendous reputation as a faith healer. He "stroked" Flamsteed in September, 1665, but the young astronomer "found not his disease to stir." When Greatrakes visited Worcester in the following February, Flamsteed went to him again, also with no result, "though several there were cured." So Flamsteed gave himself up to a life of illness and work.

It was not long before his talents attracted notice. Friends lent him books, and he repaid them by writing papers. His first published observation was of the solar eclipse of October, 1668. He accompanied it with the statement "that the tables differed very much from the heavens." The correction of the tables became his chief object and the greatest work of his life.

Towards the end of the year 1669, John Flamsteed sent a paper to the Royal Society on some calculations of appulses of the moon to fixed stars. The paper was published in *Philosophical Transactions,* and brought Flamsteed immediate correspondence. The principal writers were Oldenburg, the natural philosopher, and John Collins, the mathematician. Collins, who from that time corresponded regularly with Flamsteed, did him a great service. The astronomer tells how, in the spring of

1670, he made " a voyage to London. Visited Mr. Oldenburg and Mr. Collins; and was by the last carried to see the Tower and Sir Jonas Moore (Master of the Ordnance), who presented me with Mr. Townley's micrometer and undertook to procure me glasses for a telescope to fit it." Sir Jonas Moore, kind at this first interview with the astronomer, was to prove a still kindlier patron.

From London Flamsteed journeyed to Cambridge, where he made the acquaintance of Isaac Newton, and entered his name at Jesus College. Next year, in October, 1671, Flamsteed began his systematic observations, and "by the assistance of Mr. Townley's curious mensurator they attained to the preciseness of 5"." He soon found out that the varying dimensions of the moon completely contradicted all the theories of lunar motion except those of Horrocks, the brilliant English astronomer who had died some thirty years before. Flamsteed's observations proved the validity of Horrocks's theory and, at the request of Newton and Oldenburg, he prepared it, with additional explanations, for publication. His next work was even more important, and had a wider effect. He wrote, in 1673, a tract on the real and apparent diameters of the planets, which gave Newton the data for the subject of the third book of his *Principia*.

A year later Flamsteed took the degree of M.A. by letters patent, intending to take orders and settle down in a small living near Derby, But a greater destiny was in store.

In the summer of 1674, he was the guest of Sir Jonas Moore at the Tower. Acting on Sir Jonas's advice, he compiled a table of tides for the king's use. He also supplied the king and the Duke Of York each with a barometer and a thermometer made from his own models, and a copy of his rules for forecasting the weather by them. Early in the next year, a " bold and indigent Frenchman," styled the Sieur de St. Pierre, put forward a scheme for finding the longitude at sea. Through the patronage of the Duchess of Portsmouth, Charles the Second's mistress, he obtained a royal commission to consider the proposal. Sir Jonas Moore got Flamsteed nominated as a member, and the astronomer at once showed that the Frenchman's plan was hopeless, until they had far more accurate knowledge of the moon's course and of the places of fixed stars. Whereat Charles II cried: " I must have them anew observed, examined, and corrected for the use of my seamen."

So, by a royal warrant of March 4, 1675, Flamsteed was made "Astronomical Observator," with directions "forthwith to apply himself with the most exact care and diligence to the rectifying of the tables of the motions of the heaven, and the places of the fixed stars, so as to find out the so much desired longitude of places for perfecting the art of navigation."

A sale of spoilt gunpowder raised £520, and this miserable sum was applied to the cost of the observatory. Sir Christopher Wren "chose the site in Greenwich Park, and the building was hastily run up, from second-hand materials, to his design. Thus was founded the Royal Observatory at Greenwich, the most important and most famous astronomical observatory in the world.

When Flamsteed entered the observatory as its first Astronomer Royal, on July 10, 1676, he found it destitute of any instrument provided by the government!

While the observatory was being built Flamsteed was ordained, and he filled his time, when waiting to enter his new domain, by observing at the Tower and at the Queen's House at Greenwich. His first problem, as Astronomer Royal, was to procure accurate instruments and expert help. Sir Jonas Moore once again came to his rescue by presenting him with an iron sextant of seven-feet radius, and two clocks by Tompion. Flamsteed himself brought three-foot quadrant and two telescopes from Derby.

His salary was the magnificent sum of £100 a year, cut by taxation to £90. For this he was expected to reform astronomy and even instruct two boys from Christ's Hospital. His official assistant was a " surly, silly labourer," available for moving the sextant. In order to buy instruments Flamsteed was obliged to take private pupils, and between 1676 and 1709 no less than 140 sat under him.

In spite of all these handicaps, in spite, too, of his continued ill health, John Flamsteed, first Astronomer Royal, achieved amazing results. All the astronomical theories and tables in use were wrong. Flamsteed set himself the colossal task of correcting them single handed. His first observation for this purpose was made in September, 1676: by 1689 he had made twenty thousand.

He could measure only intermutual distances and fix the relative places of stars, for as yet he had no instrument to determine the position of the equinox. It was not until he had been presented to the living of Burstow, Surrey, in 1684, and had inherited money from his father in 1688, that he was able, with the aid of Abraham Sharp, to construct the mural arc with which his most valuable work was done. His original method, by which he determined the position of the equinox, has been called the basis of modern astronomy.

His observations on the great comet, during the early months of 1681, were transmitted to Newton, who used them in the *Principia*. The comet caused the first breach between the two scientists". Flamsteed said that it was one which had appeared in November, 1680, but Newton held that there were two comets and cast "*magisterial ridicule*" on Flamsteed's theory. But Flamsteed was right, and four years later Newton acknowledged his mistake.

Newton was now urging Flamsteed to publish a catalogue of leading stars; but Flamsteed had large schemes in view and was not interested in partial publication. This tended to widen the breach between Flamsteed and Newton, but in 1694 and l695 Flamsteed was supplying Newton with lunar observations. Here again was a source of mutual irritation. Flamsteed was often ill and kept Newton waiting, and often he was annoyed, and kept him waiting for that reason too. Newton offered him money for his help but the offer was brusquely rejected.

In 1704 Newton dined at the Royal Observatory: his aim was to find out about the catalogue which Flamsteed, to " *obviate clamour*," had said was nearly ready for printing. It was finished, and Newton offered to recommend its publication to Prince George of Denmark. Flamsteed *"civilly refused."* "Plainly," he remarked afterwards, "his design was to get the honour of all my pains to himself."

But in spite of Flamsteed, a committee of the Royal Society, consisting of Newton, Wren, Arbuthnot and Gregory, was appointed by Prince George to supervise the work, and arrangements were made for publication. The Prince was to pay. The result was deadlock, delay and exasperation, but at last, in 1707, the first volume, containing the sextant observations for 1676-1689 was published. The second volume caused still more bitterness and quarrelling. The committee went ahead without Flamsteed, the editor being his *bête noir,* Edmund Halley, who was to succeed him as Astronomer Royal.

In 1711 Flamsteed was summoned to meet the president (Newton) and other members of the board at the Royal Society rooms in Crane Court. He was asked the condition of his instruments and replied that they were his own, and that he would suffer no one to concern himself with them. Newton burst out : As good have no observatory as no instruments ! "

"I proceeded from this to tell Sir Isaac (who was fired)," writes Flamsteed, "that I thought it the business of the Society to encourage my labours, and not to make me uneasy for them, and that by their clandestine proceedings I was robbed of the fruits of my labours; that I had expended above £ 2,000 on instruments and assistance. At this the impetuous man grew outrageous, and said: 'We are, then, the robbers of your labours.' I answered I was sorry they acknowledged themselves to be so. After this, all he said was in a rage. He called me many hard names—*puppy* was the most innocent of them. I only told him to keep his temper, restrain his passion, and thanked him as often as he gave me ill names."

Such was the state of affairs between England's two greatest scientists.

In the next year the *Historia Coelestis* was published. Flamsteed's observations were incompletely and inaccurately given; Halley's preface was offensive. Halley boasted in Child's Coffee House of the care he had taken in correcting the faults. Flamsteed called him "a lazy and malicious thief, who has very effectually spoiled my work."

Flamsteed's aim now was to publish, independently, a complete and proper account of his work. He was racked by gout, by headaches and painful disease, but nothing could curb his energy. "Though I grow daily feebler,"'he wrote, " yet I have strength enough to carry on my business strenuously."

Newton acted meanly towards him; but Flamsteed had some slight revenge when, in 1714, he obtained three hundred copies out of the edition of four hundred of the *Historia Coelestis* and burned them.

He was still working and observing when he was taken ill on December 27, 1719; on the last day of the year he was dead.

Joseph Crosthwait, his assistant, and Abraham Sharp edited his works. The *Historia Coelestis Britannica,* published in 1725, contained the British Catalogue of 2,935 stars observed at Greenwich, "one of the proudest productions of the Royal Observatory." For John Flamsteed, 'the chuckling, rheumatic, crippled, suffering, irritable old man, who first studied the heavens from a bed of pain, had made one of the richest contributions to practical astronomy that the world has seen.

As for the man, he lived in his work. But he was no cold, dry-as-dust mathematician. He was a "humourist and of warm passions." He loved a joke, practical or otherwise. But he was irritable, a pardonable fault considering his almost constant pain, and his quick temper did not aid his relations with his fellow scientists. He was jealous, too, of his professional reputation, and did not like criticism or rivalry.

Against these blemishes set the facts that he was patient in suffering, pious and conscientious, abstemious and straightforward and – more important than them all – that his wife, his assistants, and his servants loved him and were devoted to him even after death.

Joseph Baron Lister

(1827–1912)

Modern hospitals and modern surgery need no encomium. The greatness of their work for mankind is equalled only by their efficiency. Yet less than a century ago our hospitals were little more than mortuaries: the work of the surgeons was negatived by the dreadful toll of poisoning that killed patients after successful operations. It was not until Lister, basing his work upon the discoveries of Pasteur, applied the first antiseptic measures, measures which are now the commonplace basis of our aseptic surgery, that this dreadful mortality was checked.

"The operation was successful, but the patient died from blood poisoning. .." How often did those ironic words appear in the bulletins of the hospitals around the world in the first half of the nineteenth century!"

The discovery of anæsthetics had so revolutionized the practice of surgery that operations, which would have been unthinkable before, were successfully carried out. Hitherto the best surgeon had been the quickest, and thus many an operation was work only half done. With the discovery and application of anæsthetics, however, all this was changed, and deep-rooted complaints were tackled and successfully eradicated. But still the word " hospital " inspired an awful dread in the public mind.

The mortality was so terrific that if only twenty-six per cent of the patients operated on in London died, the hospitals thought that they were to be congratulated. Public feeling was not appeased by this complacency, and the agitation reached such a pitch that there were many who even advocated the total abolition of the hospital system.

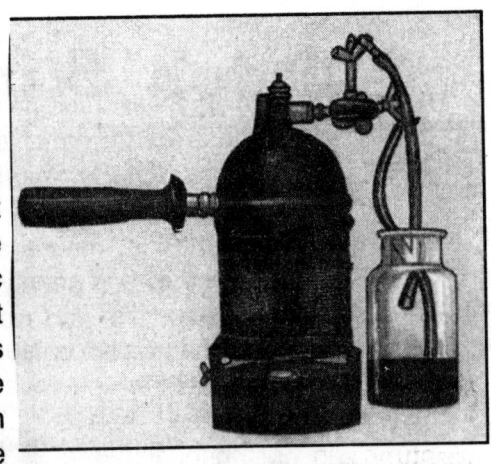

Frenzied efforts were made to meet the scourge of blood poisoning. The hospitals were swept clean of all dust. Windows were flung wide open, and stacks of clean towels and sheets were provided daily. But still the hand of Death lay heavy on the wards.

All this was to be swept away by the work of a young man whose name was as yet unknown to medical science. Joseph Lister, the son of Joseph Jackson Lister, who had been elected a Fellow of the Royal Society for his magnificent work in perfecting the microscope, was born at Upton Lane, West Ham on April 5, 1827. As his family was Quackers, he could not go to any of the schools which might have assured him a career. He was educated privately and strictly, and grew up to be a young man of serious countenance and unobtrusive dress.

In spite of the extreme strictness of his upbringing, Lister was devoted to his parents, and his father was his closest confidant and friend. He wrote to his father regularly all his life, telling him of his experiments and researches. Even after he had left the Society of Friends, he still used the "thee and thou" in his letters to his father. The old man, proud though he was of his son's success, still wrote to him in the same sober strain that he had used when Joseph was a boy.

There was nothing in this young boy when he first went up to University College Hospital in 1848 to suggest future greatness and world-wide renown. Of average height, but slender built, he had a large intellectual head, dark, luxuriant hair, and side-whiskers. He wore the curious black coat favoured by the Quackers, and altogether presented a rather odd appearance. Nevertheless, he was of a gay nature, and this was thrown into greater relief by his sincerity and gravity on other occasions. He was fond of open-air exercise, and was a powerful swimmer, and he rapidly gained the esteem of his young colleagues.

After gaining his M.B. degree at University College, London, he went to Edinburgh, where he worked under the famous surgeon, James Syme.

Even then, though invested with the dignity of "Doctor," he was not beyond indulging in student " rags," and on one occasion he went on a raiding party to take down the wooden board of a quack doctor, and burn it ceremoniously in front of the infirmary.

Another time he climbed the "Cat's Nick," was thrown to the bottom of that perilous height by his more exuberant companions, and was brought back injured to the infirmary. "I aye kent something would happen wi' you Englishmen whustlin' on the Sabbath Day," was the only comfort that he got from Dr. Syme's famous nurse, Mrs. Porter.

Syme was not an easy man to work for, as he demanded more than was usually to be expected from his young doctors. He soon saw, however, that behind Lister's quiet manner there was great application and that no detail, however small, escaped him. It was not long before Syme appointed Lister his house surgeon, and it was then that Lister performed an operation that was to live in his memory and act as a spur to many of his future achievements.

A man had been dangerously stabbed in the neck in a brawl and the wound was so deep and so close to vital veins that it was doubtful if he would survive an operation. The danger lay chiefly in the patient's weakness through the amount of blood that he had lost, and also in the amount that he would inevitably lose in the course of the operation. This case had its importance for Lister not only because it was his first operation, but also because two lives depended on the result. If the man died, his assailant would certainly be hanged for murder. The interest in the case was enormous. The amphitheatre was thronged with students, while two police officers waited outside to learn the result. Syme and Lister got ready to perform the operation. The patient was wheeled in, and the ghastly pallor of his face made everybody despair of success.

The two doctors set to work as soon as the anæsthetic had taken effect. Syme worked as if he were conducting an autopsy, coolly but with the quick precision for which he was noted. Lister, on the other hand, was bathed in perspiration, as if he had been running a race, but his fingers were no less sure than those of his senior. All his life Lister could not help sweating when he operated, and it was the only outward sign of his intense interest in his work. At last the final stitches were made. A returning colour to the patient's face told the spectators that the operation was successful. Their relief was so great that they could not restrain themselves from giving loud and prolonged applause to the two doctors who had thus saved the lives of two wretched men.

Lister's association with Syme was thereafter to be much closer, and it was not long before he met and fell in love with Agnes, the great surgeon's eldest daughter. They became engaged and were married on April 23, 1856, in the Episcopalian Church. By this act of " marrying out " as the Quaker phrase has it, Lister joined the Church of England, of which he was to be a faithful member to the end of his days. There was the closest sympathy between Lister and his young wife and, indeed, their marriage can be described as a life-long honeymoon. Agnes Lister proved to be his best and closest assistant, and she would take notes for as much as seven to eight hours a day to his dictation. Most of the notes of Lister's lectures that are extant are in his wife's handwriting.

Skilful surgeon though he was, Lister was still unable to prevent the terrible loss of life from the after effects of operations, and he devoted his every spare moment to the study and elucidation of the diseases of the blood and the causes of inflammation. His first researches were in regard to inflammation and, as a result of a paper that he read to the Royal Society on "*The Early Stages of Inflammation,*" he was elected at the age of thirty-two a Fellow of the Royal Society. From then on he became a marked man and his theories, if not always accepted, were at all events received with respect.

In 1864 Pasteur announced to the world his germ theory, which was that infection is due to germs, and that all fermentation and putrefaction come from these minute living organisms. Curiously enough, some time passed before Lister noticed Pasteur's discovery. He had been too preoccupied with his own researches into the composition of dead tissues to read much of what was being said and written outside Great Britain. When he did discover it, however, he knew that he was on the trail of the enemy that had so constantly eluded him.

The prevailing theory had always been that germs existed in the air only, and Pasteur's proof that they are to be found everywhere, even in the body, made Lister devote his attention to the state of the wounds themselves, rather than to the state of the surrounding atmosphere. Hitherto Lister had insisted on cleansing the parts round the wound and in eliminating as much as possible the entry of any foreign bodies. To achieve this end he used to spray the affected parts with carbolic acid the whole time that he was operating, and he was the first to insist on the use of clean towels and the constant washing of their hands by his assistant surgeons. But even so, patients died like flies from septicæmia. Pasteur's discovery showed Lister that a body could be infected before it had come into contact with the outside air, and he therefore set to work to discover some means whereby the germs should be killed before more damage was done.

He spent months experimenting in order to get the best results, and he finally reached the conclusion that the only way to arrest the spread of sepsis was to keep the wound in constant touch with some disinfectant which would make it impossible for any germs to live. He thus applied dressings soaked in carbolic acid—or German creosote, as it was called in those days.

Lister finally chose carbolic acid as his antiseptic because its efficacy had been proved at Carlisle where it had been used as a disinfectant for sewage. It was in March, 1865, that he first used the acid in treating a case of compound fracture. Compound fractures were commonly aggravated by the poisoning of the flesh wounds in those days and formed one of the hardest class of case with which surgeons had to deal. Lister's treatment was successful, but it was not until two years later, after he had cured a difficult case of compound fracture with no resulting suppuration and no general illness, that he published his discovery in the Lancet.

His method was to introduce German creosote, which was in fact a crude form of carbolic acid, into the wound and then cover it with a layer of lint soaked in the acid. In order to alleviate the pain made by the application of the acid, he made many experiments in antiseptics introducing one after the other, carbolic oil, carbolized putty and carbolized shellac. He also experimented with types of dressings.

These measures brought a success he had scarcely dreamed of and, in a short time, a powerful body of opinion grew up in the medical world which attached itself fanatically and almost blindly to his methods. Any small deviation from the exact procedure adopted by Lister was condemned in heated controversies which were not always too scrupulously conducted.

But Lister was indifferent to all controversies. To him it was the result that mattered, and if experiment showed that he had previously been proceeding along wrong lines, he unhesitatingly abandoned his previous method. Thus, although he was the first to use the spray in the treatment of wounds, as soon as he discovered that the danger lay not so much in the elements of the air as in the tissues of the body itself, he abandoned the spray for a more direct and efficacious method. He soon insisted on the sterilization of all surgical instruments — a measure which today seems to be elementary — but which was greeted with derision by many in the medical profession.

Lister's method of sterilization was in the beginning almost entirely by the use of acids, and it was many years before he realized the aseptic properties of heat. When he did, he immediately applied the methods that are now commonplace in every hospital, and again the controversy broke out.

His most striking success was at Glasgow Infirmary. In the same building that stands today in Glasgow the mortality was appalling; nearly eighty per cent of the patients operated on died of septicæmia and gangrene and counsels of despair were beginning to prevail. Some even advocated the destruction of hospitals where gangrene had once got a hold and it was perhaps providential that when Lister went to Glasgow he entered an institution whose *death-roll* was the highest in the kingdom. The hospital was in the most insanitary quarter of the city, and the ward that Lister had to look after was the worst of all.

He began by taking a firm line with the managers of the hospital. He steadfastly refused to allow any more beds in his ward than had been arranged for in the original scheme and it was as Lister said himself " fairly attributable to the firmness of my resistance in this matter, that though my patients suffered from the evils alluded to in a way that was sickening and often heart-rending, so as to make me sometimes feel it a questionable privilege to be connected with the institution, yet none of my wards ever assumed the frightful condition which showed itself in other parts of the building, making it necessary to shut them up for a time."

In spite of the appalling conditions in the other wards, during the first nine months that Lister practised his antiseptic system he did not have a single case in his ward of pyæmia, erysipelas or hospital gangrene. No greater proof was needed for the efficacy of his methods.

Besides being a great surgeon, Lister was above all a pre-eminent bacteriologist. Perhaps he inherited this interest from his father who had done so much to make the microscope the accurate instrument that it is today. His mind was as accurate as a machine, and the smallest detail never escaped his memory. Years after, when working on some new discovery, he would remember the result of an experiment he had carried out when a young man and put it to good use.

If he had a fault it was unpunctuality. He frequently kept whole assembly rooms waiting for more than half an hour before he arrived to deliver his lecture. This was not due in any way to laziness, but simply to the fact that he had probably thought of some new idea on his way to the lecture room which he would have to go back home to verify. The outstanding feature of Lister's work is that he never delivered a judgment which he could not substantiate with proofs. He even kept patients waiting for the same reason.

On one occasion when he had to go to a consultation on an urgent case of blood poisoning, an idea occurred to him when he was on his way, regarding a certain process to do with the coagulation of blood. Luckily he happened to be passing a slaughter house. He immediately went in and ordered a calf to be killed. Then and there, he proceeded to make his investigations. He was more than two hours late for the consultation, but he saved the patient's life.

His greatest discovery was, of course, antiseptics and their use, but he is also responsible for the perfection of the methods of stitching wounds. Hitherto very fine silk had been used for ligatures and this often caused poisoning to set in. It was Lister, who, after experimenting with animals, introduced catgut as a ligature which could be absorbed by the tissues of the body.

Such innovations and epoch-making discoveries associated with his name, have obscured the fact that Lister was a magnificent surgeon. He devised many new operations, which would have made the reputation of lesser men. And his work for antisepsis has alone made possible the surgical treatment of diseased deep-seated organs in the body.

Nothing ever ruffled this remarkable man, and perhaps the only occasion when he ever felt nervous was on the occasion of Pasteur's jubilee in 1892, when he was greeted by the assembled scientists and students with the rhythmical applause that was the highest expression of esteem ever accorded by French University custom.

In 1883, a baronetcy was conferred upon him, but he valued this honour far less than the presidency of the Royal Society to which he was elected in 1894. In 1891 he founded the British Institute of Preventive Medicine on the lines of the Pasteur Institute, and the name has since been changed to the Lister Institute. This legacy of Lister's is perhaps among the most precious that we have. Here experiments and researches are constantly being carried on to discover the causes of disease, and it is the Lister Institute that examines and pronounces on specimens of germs that are sent in by doctors from all over the world.

By nature extremely reserved, Lister never hesitated to acknowledge the sources whence he derived his discoveries, and his recognition of the debt that he owed to Pasteur is the most touching tribute that has ever been paid by one man of science to another. In 1897 he was raised to the peerage and in 1902 became one of the twelve original recipients of the Order of Merit, but these were empty honours: for in 1893, his beloved wife and companion had died, and from that time on, life held no more joy for this great benefactor of humanity.

In 1903 he had a serious illness from which he never really recovered, and for the remaining nine years of his life he was practically a cripple. He never wrote a book because his life was so crowded with work, and when his days of leisure came, Lady Lister was not there to inspire him and help him as in the past. His faculties gradually began to fail him, and like a tired child, he fell asleep on February 10, 1912, at the great age of eighty-five. He had found the hospitals of England little better than mortuaries, and he made them the great healing centres that we know today.

His reputation was so great that public opinion clamoured for his burial in Westminster Abbey. But Lister had left in his will a record — if such a thing were needed — of his devotion to his wife, and stipulated that he should find his last resting place by the side of her who had been his constant comfort and helper in the difficulties he had had to face in life.

The funeral, however, took place in the Abbey, and the great building was thronged not only by those bearing the highest names in the land but also by the poor and the halt and the lame, whose sufferings he had soothed and whose lives he had saved.

Kamala Sohonie

First Lady of Indian Science

The year was 1933.

Twenty-two-year-old Kamala Bhagwat was travelling from Bombay to Bangalore with her father, Narayanrao Bhagwat. He was obviously a man of vision and ahead of his time — inspiring his daughters, Kamala and her sister Durga (the noted Marathi literateur) to forge identities of their own.

Storming Male Bastions

Indeed, at a time when women made such journeys only for marriage proposals, Kamala had topped her B.Sc. chemistry first class from the University of Bombay and applied to the Indian Institute of Science at Bangalore for her post graduation. Father and daughter were on their way to meet the Institute director, Sir C.V. Raman, who had rejected Kamala's application.

The world renowned physicist and Nobel Prize winner told them clearly that there was no place in the Institute for female students and Kamala could very well take herself back to Mumbai. He believed that girls required "weary persistence" and that he himself was fortunate to have only sons.

No argument Kamala's father offered seemed to make him change his mind, till she firmly told the scientist that she was more qualified than many of the male students he had admitted and that to be rejected on grounds of sex was unjust. She vowed to perform 'satyagraha' before his office till she was given admission. Startled but impressed, Dr. Raman gave her admission into the biochemistry stream, though on a year's probation — if she worked hard and illustriously in that period, she would be entitled to admission like any other male student.

Kamala accepted the condition. She was able to get a lecturer, Mr. Srinivasaiya, to guide her, but again on certain conditions — that she presented herself at the laboratory at five o'clock each morning, worked diligently from then to 10 in the night, and read in the library at night. She agreed if she was allowed two hours every evening to play tennis.

By the end of her probation year — in which she published two papers — Sir C.V. Raman was so pleased with her dedication to research that he decided to throw the Institute's doors open to two more female students, allowing Kamala to complete her research and take on the M.Sc. degree course. She became the first woman to win scholarships for higher education abroad from the University of Bombay — the Sir Mangaldas Nathubhai Foreign Scholarship and the Springer Research Scholarship.

In 1937, Kamala won admission to the Sir William Dun Institute of Biochemistry at Cambridge University, England, as well as the Travelling Fellowship of the International Federation of University Women, U.S.A., in 1938.

A Woman Scientist

In 1939, she discovered the presence of cytochrome-C in plant tissue — and submitted a 40-page thesis to Cambridge University, becoming the first Indian woman to obtain her Ph.D. from Cambridge — and what's more, in just 14 months!

When Dr. Kamala Bhagwat returned to India in September '39, she had already been appointed head of the newly started bio-chemistry department at Delhi's Lady Hardinge College. Since the job did not involve research and the teaching was basic, however, she found her talent and knowledge underutilised. She took on the post of assistant director in the nutrition research laboratory of the Indian Council of Medical Research (ICMR) Institute, and suddenly found herself deeply interested in the science of nutrition.

In 1943, she made an important discovery — of the presence of vitamin P in green gram. Also, that when vitamin P (found in pressure cooked gram) and vitamin C (in the form of lemon juice) are consumed together, they help to toughen the covering around blood vessels, and thus staunch haemorrhaging. (This discovery was made next in Russia only in 1977!)

In 1947, she became Dr. Kamala Sohonie when she married Madhavrao Sohonie. In the same year, she took over as head of the biochemistry department at Mumbai's Institute of Science. From 1949 to 1969, she worked there on trypsin inhibitors in Indian field bean and double bean, and other foods. She has to her credit around 155 research papers in national as well as international journals.

A Varied Life

Dr. Sohonie founded bio-chemistry departments at Delhi, Baroda and Mumbai, and was president of the Consumer Guidance Society of India.

She also won many tennis tournaments and took part in the struggle for freedom. After devoting herself to scientific research for 36 years, she retired in 1969, but went on to experiment in her kitchen — using her knowledge in nutrition to write recipes and articles on nutritious food and food adulteration. Her book, 'Aahar-Gatha', in Marathi, was published in 1995.

She had been honoured with the Rashtrapati award, and been commended by the Indian Government for her research work — although she always rued that that valuable body of knowledge had not been implemented practically.

In June 1997, she was felicitated at the silver jubilee celebrations of Mumbai's Indian Women Scientists Association. Dr. Satyavati, ICMR president, invited her to Delhi to be felicitated at a function to be organised by the ICMR on the occasion of the fiftieth year of Independence. There, the 86-year-old pioneering woman scientist, thanked those present, and hoped her research would be used for the betterment of the nation.

Her unexpected collapse, later at the function, led to a 15-day coma, and she passed away on September 8, 1997.

Her life remains a source of inspiration to countless other women striving to live life on their own terms.

Kolachala Seeta Ramayya, Dr.

He Added Fuel to the World War II Fire

Very rarely does one come across a gripping biography of a scientist. This story is all the more interesting, since it's about a scientist of Indian origin not widely known in India — the so-called unsung hero.

The story of Dr. Kolachala Seeta Ramayya has all the necessary ingredients of a thriller. He was born into the family of an orthodox priest in Andhra Pradesh, went to America for higher studies, fell in love with an American girl, 'defected' to the Soviet Union, raised a Russian family with a girl of German blood. He developed kerosene fuel that kept the Soviet and allied forces' tanks moving in the battlefield. *For the scientific world, he became the father of Chemmotology—the science of motor oils, lubricants and additives.*

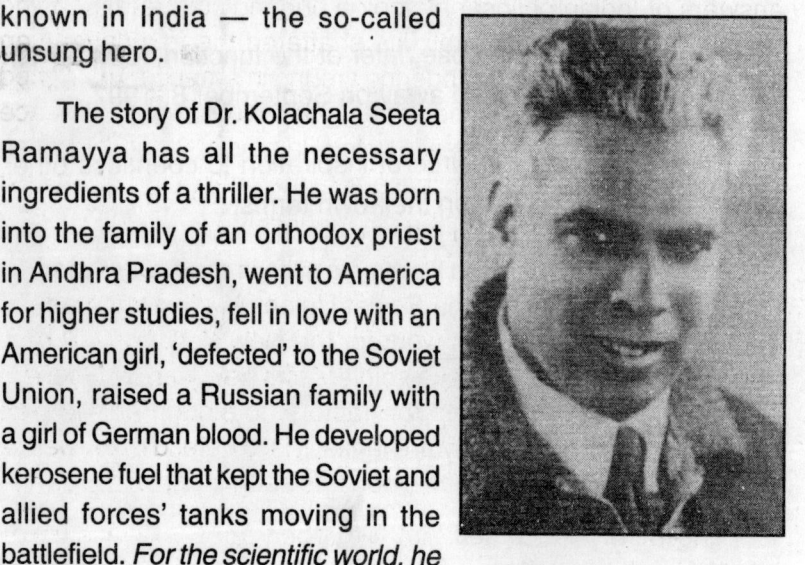

Ramayya's story is indeed a romantic story of an Indian whose inventions and discoveries in America and the Soviet Union keep the wheels of the world moving smooth. In his own words, "I did not know that a thin film of oil would grip my attention all my life, that I would see the whole world through the processes that take place in the narrow gap between the axle and the wheel."

It was Ramayya's extraordinary intelligence that earned him admission into the University of Chicago and a Master's degree in Chemistry and then a lucrative job in a New York firm that was executing contracts for the US Defence Department. It was here that Ramayya got his first patent in the field of petrochemistry in 1933. This was beginning of a brilliant scientific career.

While engaged in his scientific work at the New York firm, Ramayya was constantly thinking and evaluating the ways of American society and life. Those were the days of depression, war and mass unemployment. "Existential questions and the answers of Indian philosophy took a burning concrete meaning for me. Just like an eggshell, life shattered the foundations on which my 'I' was based. Wandering around in America, I searched for a human. I craved for the meeting that would give me peace and hope. I began losing myself," noted Ramayya. Slowly he began drifting towards communism, became a member of a local Marxist Circle and studied the Communist Manifesto.

A turning point came in Ramayya's life in the 1930s when he finally decided to shift to the Soviet Union. A fellow comrade in the US convinced that "his work for the New York firm will help making motors which will take soldiers to a new war. But there in Russia, motors are required for tractors, to plough and to take the harvest. And even if they are tanks, it is to defend in the peace for the sake of which so much blood has been shed on the earth".

The USSR welcomed him with open arms. He was made head of two laboratories — one at the oil institute and the other at the tractor institute — and was provided with all facilities to carry on his research. Here, Ramayya developed the kerosene fuel for battle tanks to operate in the USSR's changing weather conditions — this proved to be a key factor in the Soviet victory over the Germans in World War II. Some of the instruments that Ramayya developed later found their way into several countries of the Soviet block.

Konstantin E. Tsiolkovsky

(1857–1935)

Father of Rocket

It was a cold winter night at Moscow. A 16-year-old boy, who was a newcomer to the city, was roaming alone in the deserted, snow-covered streets. He was deaf, alone, hungry and shivering with cold due to lack of woollens. But his dreamy eyes were always up in the sky, looking at the twinkling stars. Suddenly, a thought struck him and he stood quiet. Was it possible to achieve this in practice? He wondered about the idea that had struck him. As a child he had tied a pebble to a string and whirled it about himself. Several times, the string had snapped, sending the pebble flying away from him. If earth was rotating, he wondered, could it also throw him away into space? Yes, man can be whirled into space and reach stars, he thought with mounting excitement. Travel to stars was possible.

Stars twinkled in the meanwhile as though beckoning him to come. With excitement he rushed home to sleep so that he could get up early in the morning and find out after consulting books whether earth could whirl man into space. The next day he met disappointment as he realised after consulting some books and doing some simple calculations that such a space travel was not possible.

Such an idea might have occurred to many young men and forgotten forever but young Konstantin E. Tsiolkovsky was not to forget it during his entire life. It occurred to him, again and again, reminding him of the promise he had then made to himself: that one day he would build something which would take man into space. In fact, his entire life was devoted to make this a reality. Unfortunately, he could not build a rocket because he was poor but he worked out all the necessary principles that make a rocket 'fly' into space. He is therefore considered as the *Father of rocket* today. His prophetic words "Mankind will not remain bound to earth" are today carved on his obelisk where he was laid to rest.

Tsiolkovsky was born on September 5, 1857, in the little town of Izhevsk of Ryazan province of the erstwhile U.S.S.R. His father was an honest and tactless man. Though he had keen interest in science and philosophy, he was a failure in life. He worked as a labourer in forest department. Young Tsiolkovsky was brought up amidst poverty because he had 12 brothers and sisters. There was however happiness in the family. His father used to enlighten him, his brothers and sisters by inventing new toys and models for them. When once his mother handed in a balloon and told him to hold the string tight lest it go up into sky, it made a deep impression on his tender mind. Those were the early stirrings of his ambition to go up into space. Unfortunately, at the age of 10, he suffered an attack of scarlet fever which left him deaf for life. His classmates teased him at school. His mother therefore took up the responsibility of teaching him. She developed his inner strength so that he could face the hazards of life boldly.

Meanwhile, Tsiolkovsky had fallen in love with books. The more he read books, the more his thirst to read increased. In fact, books and his own thoughts were his only friends. The world was otherwise silent to him. He had no friends, none to talk to and none to play game with. From a book on physics, he acquired the entire knowledge about balloons. He himself also experimented on them. Emulating his father he also built models. For instance, he built a lathe, a steam engine, a carriage which moved according to wind, etc. He even tried flying by means of wings. He could easily prove a theorem on his own rather than look for its proof in a text-book. When he was 16 he had read almost all books on scientific and technical matters available in the library of his small town. His father therefore realised that he should send his son to Moscow so that he could read the latest books on the subject. As he himself had been a victim of circumstances so he did not want that his son should also suffer the same way. He therefore decided to send his son to Moscow, come what may. All his children were also prepared to suffer for Tsiolkovsky because sending him to Moscow meant less food for them.

So, at the age of 16, Tsiolkovsky, a country boy with country manners, reached Moscow with the aim of increasing his knowledge and intellect. His father could manage to send him enough money to share a room with an old couple and eat black bread. From morning till evening he used to sit in a library, read books of his interest and perform calculations. His vision of space travel as narrated above further gripped him. Soon, he realised he should learn higher mathematics if he wanted to solve problems of space travel. He took up serious subjects such as physics, calculus and mechanics. It is said that one day when he was relaxing in a park he saw a cart packed with youngsters stop near him. He observed that as and when a youngster jumped out of the rear of the cart, it moved ahead. It then struck him that Newton's third law of motion which states that to every action there is an equal and opposite reaction, which was why the cart moved ahead when a youngster jumped out, could be used to drive a rocket in the vacuum of space. He called such a rocket "*Reaction rocket*" but he had no money to conduct *experiments* to test his idea, although in those days the Russian army had rockets. He therefore decided to half-starve himself and to utilise the saved money for his experiments! In a short while, he became bony and sick. When his father came to know this, he called him back.

In due course, Tsiolkovsky passed an examination and managed to secure a licence to teach in a school in 1879. He was appointed a teacher at a school at Borovsk in Kaluga province. Now he knew he could perform his experiments because he himself earned something. He gained respect and affection from students when they found him gentle and fair. His house soon became meeting place for students where he used to show them wonders of science by building new models and trying new experiments. There was a machine that produced electric sparks, an electric bell that mysteriously rang, a model in the shape of a hawk that flew, so on. Meanwhile, his mind was always preoccupied with space travel. Although he read quite a lot he was never systematic because he read and studied only those subjects of direct concern to space travel. There were therefore big gaps in his knowledge. He himself also knew this but he never cared because he was obsessed with space travel. An incident however occurred which embarrassed him. But it was a blessing in disguise.

In 1880, Tsiolkovsky sent three papers based on his research work to the Society of Physics and Chemistry at St. Petersburg (now Leningrad). One of the papers on the kinetic theory of gases caused considerable amusement and surprise among the eminent judges because the theory developed in it had been known for 20 years. The scientists wondered whether this school teacher had cracked a joke on them but the detailed calculations assured them that it was not so. Moreover, the other two papers on radiation of stars and mechanism of animal's organism were original pieces which impressed the scientists on the board. Among the scientists was Dimitri Ivanovich Mendeleyev, the discoverer of the Periodic Table of Elements, who immediately recognised the genius in Tsiolkovsky and made him a member of the Society, a big honour in those days. Although Tsiolkovsky was disappointed when he came to know that he had simply re-discovered something already well known and had wasted his time and energy, he also felt reassured. He felt himself capable of producing something worthwhile. Regarding the honour of membership, Tsiolkovsky felt embarrassed because he was not sure how he would look among educated city people. Moreover, he could not afford the membership fee of the Society. He had by then a family to support.

At this stage, Tsiolkovsky became fervently interested in metallic balloons — his fascination from childhood, because he thought they could be an intermediary step into space. He worked for two years without interruption on metallic balloons working out every possible details he could imagine and testing everything by his clear logic in absence of any experimental facilities. His notebooks were his laboratory and his logic his testing ground. In 1887 he was invited to give a public lecture on manned metallic balloons at Polytechnical Museum, Moscow. Eminent scientists present among the audience praised him sky high for his original thinking. Tsiolkovsky, who had been working alone without exchanging his thoughts with any one else, felt so jubilant by the praise showered on him that he fell ill ! He even lost his voice. He began to wonder what would happen to his family, if he did not regain it. Another blow struck his when his books and models were burnt to ashes in a fire. His only consolation was that the copy of the lecture on metallic balloons delivered at the Museum, an effort of several years, lay safe in the hands of its Incharge. Fortunately, he recovered from illness and also regained his voice. Soon, he was back to his thoughts and calculations on space travel.

Tsiolkovsky was transferred to a school at Kaluga in 1892. It was an industrial town with railways, factories and bustling trade, 170 kilometres south-east of Moscow. Here he began seriously his work on monoplane, a two-winged vehicle with a streamlined fuselage to carry passengers. He gave detailed designs not only of the structure of wings but also of their leading edges. He also incorporated into it his 11 year old idea of what is today known as "Jet propulsion" — the use of rockets or jets to move in air. However, he soon realised he could not be sure about his design unless he could test it in real conditions. But, how could he do that? he wondered. After all, it was then not possible to fly and determine how a design behaved in moving air? Eventually, he hit upon the idea of a wind tunnel — allow the air to flow across the design under test. But he needed a lot of money to build a wind tunnel. Fortunately, the St. Petersburg Academy of Sciences came to his aid and gave him some money to build the tunnel. Some well-wishers, who had in the meanwhile come to know about this "wonder school teacher who performed research by eating black bread" through articles on him in various newspapers, also donated money for his experiments. He eventually built the wind tunnel in his two storey house — the first of its kind in entire Russia. It was an oblong wooden box lined with metal having slits through which air was pushed in by means of bellow. For the first time, Tsiolkovsky was able to test how his designs fared in flowing air conditions. After two years of strenuous labour, he wrote a paper *"The Airplane or a Bird-like (Aviation) Flying Machine"* ! Though it was widely discussed he was not at all recognised for his work by the scientists who mattered.

Tsiolkovsky's real breakthrough in the field of rocketry came in 1903 when his paper *"Exploration of Outer Space by Rocket Devices"* appeared in a prestigious journal *Scientific Review* after it was held up for publications for five years. This 50 page paper gave a complete outline of the principles of the rocket that would take man into space. It was the result of 20 years of obsession with spaceflights. But it was a masterpiece blueprint of modern rocket. Without firing a single rocket, he had given not only the principles of rocket flights but also its design, components, materials and fuels in details based on available scientific and technological information. In the paper he began with the definition of outer space and claimed that it was nothing but vacuum, that air was present only upto 56 to 64 kilometres above the surface of earth. In vacuum he emphatically claimed that no engine could work except reaction rocket or engine. He believed that his rocket would mainly compose of a long metallic combustion chamber, fuel tanks, a fuel pump and a sealed cabin with oxygen supply for the crew. Tsiolkovsky also added that the powder rockets then in use in military were not suitable for space travel because the fuel was heavy. He suggested instead the use of liquids propellants. After much indepth study of all rival fuels such as kerosene, methane, pertrol, etc, he preferred the use of oxygen and hydrogen liquids as propellants to drive the rocket. The two liquids would be kept separately and then mixed and burnt to produce the necessary motive force to drive the rocket. Tsiolkovsky also discussed stability problems while a rocket is in space and the hazards faced by it while it re-enters the atmosphere of earth. Even he had forwarded the concept of multi-stage rockets, a chain of rockets firing one after another until the last rocket escapes the gravity of earth. It was a thoroughly worked out blueprint which the Soviet space scientists simply converted into a concrete reality. But when Tsiolkovsky's paper appeared, nobody commented upon it. Moreover, the journal also closed down soon after. Tsiolkovsky's blueprint therefore remained lying in libraries gathering dust until he was recognised by the Soviet Communist party which came into power in 1918.

Undaunted by the failure of his paper to evoke proper response from scientists, Tsiolkovsky meanwhile began to write about his ideas on space and space travel in newspapers. He wrote as if he was right in space. He discussed the problems of launching into space, the kind of spacesuit one had to wear, the kind of space conditions present, so on, so forth. He also predicted the installation of technical stations — satellites — in space and also space cities. He also showed how artificial gravity conditions could be created in space. In short, he realised that one day earth would be overpopulated and mankind would therefore need extra space to settle. He also talked of harnessing solar energy for food production and energy generation in space. The question of communication with alien life on other worlds was also uppermost in his mind and he discussed it at length giving his own ideas on the subject. To popularise his ideas on space, he also wrote a science fiction *Beyond the Earth*. In the meanwhile, one populariser of science Yakov I. Perelman had also begun to popularise Tsiolkovsky's ideas on space travel. The result of all these activities was that in 1929, a society for space travel was founded at Moscow. It was called the *"Group for the study of Reactive Motion"*. In the meanwhile, Tsiolkovsky was also hailed by the new Soviet Government as the *Father of Rocket*. He became a national hero overnight. His 75th birthday anniversary was celebrated with pomp and show all over Russia in 1932.

In his later years, the Soviet Government gave Tsiolkovsky pension, so that he could work peacefully without worrying about food. He was also awarded the *"Red Banner of Labour"* three years before his death on September 19, 1935. He died only after having a glimpse of stars through his dim eyes. If space travel did not become a reality during his life, it certainly became soon after 100 years of his birth — the only prophesy that he could not make. The first satellite Sputnik-I was launched into space in October 1957 and opened the doors to the present Space Age. As a mark of honour to him, the Russian (earlier U.S.S.R.) Academy of Sciences awards Tsiolkovsky Gold Medal every year to any space scientist in the world who has made an outstanding contribution the the field of spaceflights.

Shuttle Era Begins

On April 14, 1981, the doors to a new era of space exploration and exploitation was opened when the U.S. National Aeronautics and Space Administration (NASA's) Space Shuttle Columbia made a perfect landing like a glider on the Dryden's dry lake, California, after climbing up into space like a rocket from Kennedy Space Center, Florida. Such a winged spacecraft the space scientists had dreamt of even in the early fifties, when *Space Age* had just begun. The reusable Shuttle is of the size of a D.C.

aeroplane and has all its looks but for the four big engines at its rear and the heat-resistant tiles that cover its body. The cockpit is just behind the nose of the shuttle and below the cockpit and the living quarters of the astronauts. Just behind the cockpit is a huge space, called cargo bay, 4.5 metres in diameter and 18.3 metres in length, which ends at the rear of the shuttle. As the name implies, cargo bay is a space in the shuttle for carrying men, materials and equipment upto a weight of 27,215 kilograms into space and back. The cargo bay had two huge curving doors that open on at the top of the shuttle, which can be opened or closed in space as required. It also had a one metre long robot arm for picking objects from or releasing objects into space, which could be handled by the astronauts in the cockpit by means of computers and TV cameras scattered around the cargo bay.

Another vital addition to the shuttle is the *Spacelab*, a multidisciplined laboratory, which is assembled by the European Space Agency. The Spacelab consists of several chambers which could be fitted into one another in any order as required, like blocks in a mechano set. Each chamber has its own speciality. One has pressurised atmosphere in which scientists work with as much convenience as available in a terrestrial research laboratory. Another chamber has equipment such as furnaces, computers, etc, for conducting various experiments in the almost zero-gravity and vacuum conditions of space. Yet another chamber has open-to-space environment for studying the various phenomena occurring in space and down on earth. The Spacelab has its own jets for manoeuvering once it is released into space by the robot arm from the cargo bay of the shuttle.

What does the shuttle do? It carries men, equipment and satellites into space at a far reduced price and faster than conventional throwaway rockets. Technological, physical and biological experiments carried out in space help in the production of new alloys, new drugs and new crystals which would break new grounds in the field of materials, pharmaceuticals and micro-electronics. The shuttle could also repair on-the-spot faulty satellites in space, which nowadays are lost forever, and bring those back home for any major repair. Of course, such a capability could be misused at the time of war to kidnap enemy satellites. With all kinds of detectors and telescopes attached to the shuttle, looking both up towards the heavens and down on earth, would, apart from studying radiations coming from outer space and in locating mineral and fossil fuel deposits, etc, also keep a watch on the enemy territory. At a future date the shuttle would also haul men and material into space to assemble a big telescope, solar energy satellites, permanent space stations and space colonies, and so in a way would help mankind in solving its present problems of energy crisis, food and population explosion.

In near future NASA intends to make shuttle trips into space available to a man of any profession, whether for his personal experiments or for a trip into space, on a small fee. The man or woman wanting to experience space should be healthy and in the age group of 25 to 65 years. On the trips into space, hot meals and ice cream have already been promised to the shuttle travellers. A space traveller has, however, to undergo three months training in a space center to qualify for the space trip. According to NASA experts every shuttle could be used for 100 trips into space and at present efforts are on to reduce refurbishing and re-fulling time between two shuttle launchs. NASA's target is to have 60 shuttle launchs into space every year. Besides NASA's shuttle, efforts are presently underway in the U.S.S.R. and Japan to build their own shuttles. No doubt the *Shuttle Era* has begun.

Mankind, however, should watch the *Shuttle Era* with caution because along with the good it also forebodes evil. Owing to the shuttle the next wars would be fought in the coldness of the space, bringing in their wake immense horrors on the surface of earth. Then the frequent launches of the shuttle into space will also poke holes in the ionosphere (the layer which saves mankind from deadly radiations coming from space) which could in the long run destroy either of these layers and prove disastrous to mankind. Let's hope the Shuttle Era proves more beneficial and less harmful to mankind than it has yet been portrayed.

Leonardo Da Vinci

(1452–1519)

The Man of Many Minds

I

In the autumn of 1517 a party of Italian visitors arrived in the small French town of Amboise, one of the favourite seats of the young king, Francois I. The party was headed by the Cardinal Luigi of Aragon, and like all visitors the Cardinal took the opportunity of seeing the local sights.

Not far from the town, in a small Gothic manor house of brick and white stone, was living one at whose presence there the King of France felt proud. On October 10 the Cardinal and his suite set out for the manor house at Cloux. Conveniently, his secretary accompanied him and later made some notes about their host of that day: "Messer Lunardo Vinci, an old Florentine...the most eminent painter of our time."

Autumn was a suitably symbolic season for their visit to Leonardo. The painter was growing old in a foreign country — he seemed more than seventy though he was some years less; he was ailing and his right hand had become paralysed; it was unlikely he would ever paint again. But he aroused himself for his fellow countrymen. He showed them some pictures he had painted before paralysis. He opened the many volumes of his manuscript notebooks which covered so many subjects : anatomy, hydraulics, geology, botany, mathematics. The Cardinal and his suite must have seen also the very small group of friends and pupils about

Leonardo. And there was Leonardo himself, still an impressive figure, with the long beard of a great magician or a Victorian father, settled at last in the security of his eight-roomed manor which he was not to quit until death.

His life had consisted of many journeys, of serving under different patrons, owing no allegiance to anything but his genius: a life lonely and self-contained, elegant without wealth, famous but not always successful. Some at least of these strands went back no doubt to his childhood without a mother and to the circumstances of his birth. He was the illegitimate son of a peasant girl, Caterina, and a successful lawyer, Ser Piero. He was brought up in the Tuscan countryside, near the small town of Vinci (whence his name), for it was in this district, at the village of Anchiano, that he was born in 1452.

II

Quite early he displayed such ability in drawing that his father showed some of his sketches to an artist friend, Andrea Verrocchio. Presumably at Verrocchio's suggestion, Leonardo entered his studio at Florence. Verrocchio's studio was a busy place, occupied not only with painting but with goldsmith's work and sculpture in bronze and stone. It was in painting, however, that Leonardo declared himself, by painting an angel in Verrocchio's *Baptism of Christ* which, so legend says — was so good that Verrocchio, sulking at his pupil's talent, never painted again. Fortunately the picture survives at Florence and from the style it is clear that the angel at the furthest left is by a different hand – that of the young Leonardo.

In these early years Leonardo began to make his mark not as a painter but as a personality. The immense glamour about his name is no recent cult and there has been no period when his genius has been neglected or forgotten. In the rather simple life of everyday Florence, where the artists divided their time between quarrels and practical jokes, Leonardo's aristocracy of spirit marked him off as different from the average artist. He was aloof, graceful, elegant and cultured. In addition to ability, he had great personal beauty. Vivid brief echoes of his character come down to us: of his brilliant conversation, his talent as a musician, his love of clothes. He would wear a short rose-coloured cloak at a time when most people wore long cloaks. He loved horses and birds — often he went and bought caged birds and freed them. All forms of life he respected, probably to the point of avoiding eating animal flesh.

This person was not clearly going to enjoy the steady work-a-day existence of the average painter or sculptor. Indeed Leonardo had left on record his scorn for sculpture and the sculptor who gets covered with sweat and dust and chips when chiselling the stone. The painter, on the other hand, he thinks of in romantic terms as painting in fine clothes, listening while he works to the sound of music or to a book being read aloud. Dreaming of some such existence, Leonardo seems to have been allured from the prosaic life of Florence by the softer air of Milan. In Florence the virtual ruler of the city and great patron of artists, Lorenzo de' Medici, does not seem ever to have employed him. And when Leonardo went off to Milan it was, an early writer tells, as a musician, carrying a marvellous silver lute of his own design shaped like a horse's head.

III

At Milan the Duke Lodovico Sforza was attempting to create out of the duchy be had seized a cultured, magnificent and impregnable city. Leonardo believed he could be of use to the Duke in all these ways and he drew up a letter of self-recommendation eminently practical and without any mention of silver lutes. In this he sets out at length his qualifications and his proposals as a military engineer, the instruments of war he can construct and the terror he can cause. Leonardo, so passionate for a bird's liberty, describes dispassionately his machine for bombardment; and this curious coldness can be detected again and again in his comments and in his drawings. He dissected a body without repugnance and drew the face of a freak apparently without pity. At the end of his letter to Lodovico Sforza Leonardo mentions, almost casually, that he hopes to be of use in a time of peace as architect, engineer, sculptor and painter.

The Duke seems to have been impressed. At the age of just over thirty, Leonardo was working in Milan. Yet the promise of the place was never quite fulfilled; Leonardo, too, was perhaps never fulfilled. His years at Milan were far from idle, but many of his occupations were trivial, others unfinished, others doomed to early decay. Already there was a Hamlet-like flaw in the centre of his activity. His desire for the best often resulted in nothing being accomplished. His dreams were divine but incoherent. His experiments both scientific and artistic fascinated him but seldom justified themselves in practice.

When he left Florence he probably left behind unfinished an altar-piece. *The Adoration of the Kings,* for some monks who patiently but vainly waited fifteen years for him to complete it. In Milan, Leonardo's first task was to paint an altar-piece, *The Madonna of the Rocks;* this time his failure to complete the picture resulted in a long and complicated lawsuit not settled for more than twenty years—during which time the picture remained half completed. There are two versions of this famous composition, in Paris and in London; the London version remains even now not quite finished.

For the Duke, Leonardo began a sculpture called 'the Horse' to be a monument to the Duke's father. Of this only the preliminary model was completed and set up. Later when the French invaded Milan their archers used it as a target. Nothing has survived of it but Leonardo's many preliminary sketches.

One task Leonardo did after delays complete: the fresco of the *Last Supper* for the refectory of the monastery of Santa Maria delle Grazie at Milan. But in this instance he experimented with the medium he used and, on the damp wall, the fresco began to deteriorate during his lifetime. Again and again it has been restored, the last time quite recently; but nothing can bring back what is lost and the wreck on the wall today is a faint shadow of Leonardo's fresco.

The prior of Santa Maria delle Grazie had a young nephew who used often to see Leonardo and years later he remembered how the painter could occasionally be seen walking the hot empty streets of Milan at midday, coming to paint—or just to gaze at—his fresco. Sometimes Leonardo would climb the scaffolding, give a few touches to the wall with his brush, then leave the building for the rest of the day. At other times he came but did no painting—simply stood considering the composition for some hours.

He aimed so high that the execution of anything never quite satisfied him. Nothing was simple, and each fact led him on to another, until he was an eternal Sherlock Holmes forever pursuing a vital clue which always eluded him. He himself was well aware of this. Across many of his drawings he wrote, "Tell me if anything was ever finished." And this refrain occurs again and again : "Tell me, tell me if ever. .." he scribbles over the paper, leaving the sentence itself unfinished.

IV

Painting was only one part of his task at the Sforza court. Like all Renaissance artists he had to be a handyman, needed as pageant master and scene designer, architect and engineer. Over a question of heating the Duchess's bathroom he was called in; over a play acted at court; over fortification of the Duke's castle. It was perhaps all this variety of job which prompted him to start his vast series of notebooks, packed with hundreds of his observations, recorded by writing and drawings. The notes are jumbled together: whole passages of books read, observations on anatomy, on clouds, on faces seen in the street, accounts of how much he has spent on food or on clothing one of his servants, drafts of letters written, plans for cities, diagrams and drawings of prodigal beauty. Better than anything else, these books map the extent of Leonardo's studies and demonstrate their colossal range.

Outwardly he showed no vulgar signs of activity. He appeared to dawdle over projects. He remained elegant and handsome and self-absorbed. His studio was full of young men — lists of their names occur among his notes — who worked for him and loved him. Some, like his favourite, Salai, behaved badly; he notes Salai's behaviour, but he fondly buys him green and silver clothes in which Salai can ape his master's exquisiteness.

Perhaps Leonardo would never have left Milan, but events compelled him. As time passed the Duke grew dilatory. The glamour of his duchy dimmed as the threat of French invasion became imminent. He had soon no time for artistic affairs. In the autumn of 1499 he was deposed and in December Leonardo carried away his genius, and his pupils, from Milan. No loyalty held him. He drifted about Northern Italy: now at Mantua, now at Venice. Then, after eighteen years absence, he turned to Florence.

V

Lorenzo the Magnificent was dead and the city had declared itself a republic. But the old passion for art remained. It was as a great artist that Leonardo returned home and soon the crowds were pressing into the room where for two days was exhibited his drawing of the *Madonna and Child with Saint Anne*. This is said to have been sent later to France. It is typical of Leonardo's refusal to be committed in political loyalties that he had already found patrons among Lodovico Sforza's enemies and after his return to Florence he was busy with a painting for the secretary of the King of France.

As usual, however, painting did not satisfy him for long. He had not been back many weeks before he lost patience with his brush. Instead he was occupied studying geometry. Then he widened his activities beyond abstract science. A friend of his was serving as captain in the army of the Pope's war-like son, Cesare Borgia. Leonardo joined the staff as military engineer.

The Borgia prince was Fortinbras to the Hamlet of Leonardo. Ambitious, audacious, practical, he had embarked upon a conquest of all Italy. Under him Leonardo was to find action at last—but action as dangerous as it was exciting. First he made maps for Cesare; then he accompanied him to besiege a city; during a rare respite from war he drew the prince's portrait. Suddenly, on the last day of the year 1502, Leonardo's friend and Cesare's captain was strangled — on Cesare's orders. Perhaps Leonardo felt some portent. Surprisingly soon he was off the staff and back in Florence where life, if duller, was safer.

Here he resumed painting, with a portrait which has become almost too famous. The second wife of an ordinary citizen Francesco del Giocondo sat for her portrait and the result was the *Mona Lisa*. Early the picture was the object of stories: how Leonardo worked at it four years and to preserve the sitter's smile had music played and jesters to perform (though the jokes must have been better than most samples of Florentine wit which have come down to us). Although the *Mona Lisa* began as a simple portrait of an ordinary woman, it changed under Leonardo's hand until it became the strangely smiling image in the Louvre today. Heavy varnish and repainting have helped the mystery and obscured a face which originally possessed—and perhaps underneath still does possess—startling lifelikeness and vivid colour.

As well as private commissions Leonardo now received a public commission from the city council. He was to paint a patriotic fresco of a Florentine victory, the *Battle of Anghiari* for a hall in the Palazzo Vecchio ; a short while afterwards his younger rival Michelangelo, was commissioned to paint a fresco opposite in the same room. The two great artists were too temperamentally contrasted for there to be any sympathy at all between them and the idea of cooping them up together in the same room was audacious. Their rivalry should have resulted in the Florentine government gaining two splendid works of art. Unfortunately neither painter ever came near finishing his fresco. Michelangelo was summoned away by the Pope. Leonardo experimented, as was his way, and the process again went wrong. He did not proceed. Then the French governor of Milan asked to borrow him and he, too, left Florence. But the Florentines pressed hard for his return, yet when he came it was only because of a law suit resulting from his father's death. He did not touch the *Battle of Anghiari,* a portion of which remained for some years on the wall and then was obliterated.

The King of France now asked that Leonardo should return to Milan; and for the French for some years he designed pageants, planned engineering feats, projected statues.

VI

Five or six years later, however, found him in Rome, where the younger son of Lorenzo the Magnificent had recently been elected Pope. Indolent, pleasure-loving, good-tempered and artistic, Leo X was the perfect patron and from everywhere Italian artists migrated to Rome to enjoy the Papal favours. Leonardo and his pupils were housed in the Vatican and it seemed that the Medici family was at last to patronise the ageing painter. The Pope's brother became his protector and the Pope himself ordered a picture.

But Rome was a centre of artistic rivalry as well as activity. Younger men, like Michelangelo and Raphael, were already employed and were already famous. Leonardo could emulate neither Michelangelo's titanic energy nor Raphael's graceful facility. His methods were complex, intricate, so pondered upon that even the Pope's good nature rebelled. "He thinks of the end before the beginning!" Leo exclaimed on hearing that Leonardo was considering a new recipe for varnish before even starting a painting. And, it seems true, that the picture was never completed—perhaps was never begun.

Leonardo could not satisfy a vulgar demand for results; the person he wished to satisfy was himself. He withdrew into isolation, playing with toys of his own invention, dabbling at various things but achieving nothing. He had a pet lizard which he equipped with false wings and horns to make out of it a dragon to frighten people; he amused himself by studying and experimenting with mirrors. One or two paintings, now lost, are recorded at this time, but probably he painted little during the three years or so of his stay in Rome. While Raphael drove himself towards premature death and Michelangelo drove himself to premature old age, Leonardo idled. Even when he tried to work there would be trivial but annoying hindrances. One of his craftsmen, a German, played truant and would not work, would not learn Italian, preferred to go out shooting with the soldiers. Leonardo drafted time-consuming letters of complaint on the whole subject to the Pope's brother. Then his patron left for France where he was to be married, and afterwards came news of his sudden death.

VII

Leonardo appeared to be deserted on every side, but a last patron, perhaps his most devoted, summoned him. This was the new king of France, François I, flashy, handsome, fond of jewels and women—and artists. Like so many Renaissance princes, he wanted the glory of having employed great men. In Rome was a genius at leisure, an old magician whose prestige in France had not lessened. And at the court of François, Leonardo would be honoured and welcomed and safely lodged for life.

He set out on one final journey, to France. The king gave him the manor of Cloux, close to Amboise where he himself often stayed. The presence of Leonardo was enough; he seems to have been employed on little, but the king came many times to talk with him, and years later said that he thought no other man knew so much about the arts as Leonardo. The conversations of the old artist and the young monarch have been lost to us. We hear however of one final fantastic spark of Leonardesque ingenuity in the machinery of a lion that he designed. At a court masque this advanced upon the king as if to attack him, and then its head opened to reveal the lilies of France upon a blue background.

Even this piece of pageantry was behind Leonardo when the Cardinal of Aragon came to see him. From the next year survives a broken sentence, written on St. John's day 1518. This is the great Florentine feast because St. John is patron of the city and Leonardo writes down, as if to note the strangeness of being so far away on such a day: "in the palace of Cloux at Amboise."

It was his last St. John's day. On May 2, 1519 Leonardo died at Cloux. Some time before he had made his will, stating the masses to be sung for his soul, the poor men who were to carry torches at his funeral. He remembered his half-brothers at Florence and a French servant, perhaps his housekeeper, at Cloux. To Francesco Melzi, his faithful Milanese pupil, he left those notebooks which contained so much of his personality.

There he had sketched solutions to problems which seemed uninteresting or ridiculous to most of his contemporaries. He nearly anticipated Harvey's discovery of the circulation of the blood. From fascination in the idea of flight evolved his plans for machines with wings so that men could fly like birds. In his beautiful backward-sloping handwriting—for he was left-handed and wrote in that way—he noted his observations on anatomy, optics, geology. Implicitly he had grasped the theory of evolution. And all these interests were illustrated by the myriad drawings—in pen, or silver pointed pencil, or crayon. This was his testament, his justification for the years of procrastination.

His loss, Melzi wrote sorrowfully, is a grief to everyone, for it is not in the power of nature to reproduce such a man.

Linus Carl Pauling

(1901–1994)

Father of Molecular Biology

Linus Carl Pauling was a renowned scientist and dedicated peace activist, and the only person to have won two unshared Noble Prizes—for chemistry in 1954 and peace in 1964.

Born in Portland, Oregon, USA, Linus was the son of a self-taught pharmacist, Herman Pauling, who lived on the brink of poverty all his life. When Linus was nine, his father died, leaving the rest to fend for themselves. Linus' mother, Isabelle, struggled with financial difficulties, chronic illness and depression for the rest of her life.

Science offered Linus respite from family struggles. By age 11, he was collecting insects; at 12, he began collecting rocks and minerals, and at 13, he became interested in chemistry, when his friend, Lloyd Jeffress, showed him how sulphuric acid could change sugar to steaming black carbon. Jeffress and Pauling built their own chemistry laboratory, using discarded equipment and chemicals, and they would annoy and amuse their neighbours by making stink bombs and loud explosions.

In high school Pauling took too many maths and science classes. He waited until his final semester to enroll for two required history classes. He wasn't allowed to take classes concurrently, and so, even though he had more than enough credits to graduate, he left high school at age 16, with no diploma. He is perhaps the most successful high school dropout in history. In 1917, after quitting high school, Pauling worked as an apprentice machinist, for $ 50 a month and dreamed of becoming a chemical engineer. He was determined to go to college despite his mother's reluctance. Pauling's sister remembered that "Linus was always thinking. His mind was just active all the time, wondering about this and that and the reasons for them."

In October 1917, Linus Pauling, age 16, arrived at Oregon Agricultural College (now Oregon State University). At 18, Pauling began teaching basic chemistry classes at the college. In his senior year, one of his students was Ava Helen Miller, "the smartest girl I ever met." They were to be married a couple of years later.

After graduating with a degree in chemical engineering, Pauling resisted pressure from his mother to stop studying. Pauling was 21 when he entered the Caltech graduate programme in physical chemistry. Professor Arthur Noyes, one of the nation's time best chemists, was head of Caltech's chemistry department. One of Noyes' primary interests was the nature of chemical bonds. One year later, Pauling married Ava Helen.

Pauling's advisor at Caltech was Roscoe Dickinson, an expert on X-ray crystallography. Pauling became one of the first American chemists to effectively use the X-ray diffraction technique. Much of his early research used this technique to measure the distance and angles of bonds in inorganic crystals including topaz, micas, sulfides and silicates. Pauling trained others in X-ray diffraction, including Willum Lipcombs, who would receive a Noble Prize in chemistry. Pauling received a PhD in chemistry with high honours in June 1925.

A year later at 25, he received a Guggenheim fellowship to study at the University of Munich under Arnold Sommerfeid, a theoretical physicist. Here he began work with quantum mechanics. In January 1927, he published "The Theoretical Prediction of the Physical Properties of Many Electron Atoms and Ions; Mole Refraction, Diamagnetic Susceptibility and Extension in Space", where he applied the concept of quantum mechanics to chemical bonding. In 1928, he published six principles to decide the structure of complicated crystals. At this time, Pauling took an assistant professorship in chemistry at CalTech. In 1928, he published a paper on orbital hybridisation and resonance.

During the 1930s, Pauling completed his most important scientific research and publications, which dealt with the nature of chemical bonds. These are attractions between atoms that hold molecules together. Atoms bond together to form molecules in a variety of ways. Pauling discovered how the nature of chemical bonds determine the structure of molecules. He had already suggested that the structure of molecules was the key to understanding their chemical properties and how they reacted with other molecules. He revolutionised the way scientists thought about chemistry by offering new ideas about the nature of the chemical bond. In 1931, he published "The Nature of the Chemical Bond", where he theorised that in order to create stronger bonds, atoms can change the shape of their waves to that of petals, also known as the hybridisation of orbitals.

That year Pauling was awarded the Langmuir Prize by the American Chemical Society, for the most noteworthy work in pure science by a man under 30. At this time, he was offered a joint full professorship in both the chemistry and physics at the Massachusetts Institute of Technology. The same year, he was made full professor at Caltech. In 1933, he was made a member of the National Academy of Sciences, the youngest ever appointed to this body at 32. He was also appointed chairman of the chemistry and chemical engineering division at CalTech. In 1939, he published his most important book, *The Nature of the Chemical Bond*.

Pauling is also considered the father of molecular biology. He wrote "My serious interest in what is now called molecular biology began in about 1934". He discovered that magnets repel haemoglobin in arteries and attract haemoglobin in veins. The work on haemoglobin led to work hydrogen-bonding between the polypeptide chains in proteins. Proteins are made of amino acids, large molecules made up of smaller molecules. Most protein molecules are made of hundreds, sometimes thousands of amino acids, joined together by peptide links into one or more chains or polypeptides.

In 1948, Pauling worked out the alpha helix structure of a polypeptide. He was at Oxford at the time, confined to bed with nephritis. He later said, "I took a sheet of paper and sketched the atoms with the bonds between them, and then folded the paper to bend one bond at the right angle, and kept doing this until I could form hydrogen bonds between one turn of the helix and the next. It took a few hours. to discover the alpha helix". In 1954, Pauling was awarded the Nobel Prize for chemistry, for research into the nature of the chemical bond and its application to the elucidation of complex substances—in other words, for the body of his work "rather than for a specific discovery , a move unprecedented in the history of the Noble.

Throughout the 50s and 60s, Pauling spoke vigorously on the perils of atomic fallout and against war. Einstein headed the Emergency Committee Of Atomic Scientists, which he invited Pauling to join in 1946. In later years, Pauling said it was "Einstein's example that inspired my wife and me to devote energy and efforts to pacifist activities." Shortly after the US dropped atomic bombs on Hiroshima and Nagasaki to end World War II, the Paulings joined other scientists to warn of the perils of fallout.

In 1958, Pauling wrote 'No More War!', which discussed the threat of nuclear war and testing. He and Ava Helen submitted an anti-nuclear petition to the UN, which included the signatures of over 11,000 scientists from 49 countries.

On 10 October 1963, President John F Kennedy and representatives from the UK and Soviet Union signed the partial nuclear test ban treaty that the Paulings had championed. On the same day, the Nobel Committee announced that Pauling had won the Nobel Peace Prize. In the presentation speech, Gunnar

Jahn, chairman of the Nobel Committee, said, "Since 1946, (Pauling) has campaigned ceaselessly, not only against the spread of these armaments, not only against their very use, but against all warfare as means of solving international conflicts".

Since his second Nobel Prize, Dr. Pauling researched the chemistry of the brain and its effect on mental illness, the cause of sickle cell anaemia and what happens to haemoglobin in the red blood cells of anaemic people, as well as the effects of large doses of vitamin C on common cold and some kinds of cancer. He also published a paper on high temperature super conductivity. He worked at the University of California at San Diego, Stanford, and the Linus Pauling Institute for Medical Research. He won 75 awards for chemistry and peace. He also received the Gandhi Peace Prize in India in 1962. In the summer of 1994, when he was 93 years old, he received a standing ovation at a symposium organised by the American Association for the Advancement of Science in his honour. He died a few weeks later.

Pauling provides a model for both young and old to pursue learning, knowledge and activism, throughout their lives. Born on the frontier, be spent his life tackling the frontiers of science and humanitarianism. His commitment to peace and his vast contribution to varied fields of science will be remembered forever.

Louis Pasteur

(1822–1895)

Science In The Service of Man

I

One July day in 1885 a little Alsatian lad named Joseph Meister was wending his way to school when suddenly a mad dog sprang out upon him. Joseph was only nine years old and too small to be able to ward off the savage attack; all he could do was to throw up his hands to protect his face. Fortunately a bricklayer saw the incident. He rushed to the boy's rescue and beat off the dog with an iron bar he was carrying. Then he gently picked up the badly mauled child and carried him to the nearest doctor's surgery.

Meantime the dog had returned to his master, the village grocer, whom he bit. Whereupon the grocer seized his gun and shot the rabid beast. Afterwards a post-mortem examination was carried out and the dog's stomach was found to be full of hay, straw and bits of wood. There was no doubt about it now: the dog was suffering from rabies, which can be transmitted to human beings, causing the dreadful disease called *hydrophobia* — in which the sufferer has a most terrible thirst but cannot bring himself to touch water.

Now at that time there was no cure for *hydrophobia*, and, as mad dogs were not uncommon then, everyone dreaded the effects of their bite. The doctor who examined the wounded boy could do little beyond washing the bites with carbolic acid. He knew that wounds like these inflicted by a mad dog were commonly fatal, and to the parents of young Joseph he could offer small hope that their son would recover. He told them there was only one chance, and that a very forlorn one. This was that the mother should take the boy to Paris with all speed and seek out a scientist called Louis Pasteur. He explained that Pasteur was not a medical man. At that time there was not a doctor in all France — nor even in the whole world—who knew how to cure rabies, but Pasteur had been experimenting with a vaccine which had worked successfully on dogs. After vaccination the dogs could be bitten by mad dogs without the usual dire results — they had become immune from *hydrophobia*. The vaccine, however, had not yet been tested on any human being.

So it was that Joseph, accompanied by his mother and by the grocer who had owned the mad dog, arrived at Pasteur's laboratory on July 6, 1885. Pasteur was shocked when he examined the lad and saw the fourteen ugly wounds on his body. He was told that two days had elapsed since little Joseph had been bitten, and Madame Meister then asked him whether he could do anything for her child. Pasteur was unable to give her an immediate answer, for he did not feel sure that he could risk treating the boy with the vaccine which had never yet been tested on a human being. He could do nothing until he had discussed the matter with Professor Vulpian of the French Government's Rabies Commission. The professor had been extremely impressed with the power of the vaccine to protect dogs and he expressed the opinion that it would be just as successful with human patients. After examining Joseph, he told Pasteur that the boy would almost certainly die unless he received the vaccine. And so the decision was taken to innoculate little Joseph Meister. With a hypodermic syringe a few drops of vaccine were injected into his side. It was a good omen that Joseph stopped crying almost immediately after the injection; he dried his tears when he realized that the slight prick from the hypodermic needle was all that he would feel with vaccination. It was not long before the child sat up and began to take notice, of his surroundings. To a child Pasteur's laboratory was paradise, and young Joseph lost little time before he was playing happily with the experimental animals—the rabbits, white mice and guinea pigs—which belonged to his friend, dear Monsieur Pasteur, as he called the scientist.

Ten days after his arrival in the laboratory Joseph was given the last of a series of fourteen inoculations. Pasteur was beginning to believe that the experiment was going to be a success. "Perhaps one of the great medical events of the century is going to take place," he wrote. By day Pasteur was optimistic; this could be the beginning of a new era in which rabies could be cured. But at night he was a prey to his worst and his sleep was troubled with visions of the child suffocating in the last throes of *hydrophobia*, like a child he had seen die of the disease some years ago.

Then came the final test—the injection which would prove whether the child was now immune against the disease. He was injected with a culture of virus strong enough to produce rabies inside seven days when it was given to a rabbit. If the boy took this injection without harm, then the treatment was complete. To the great joy of everyone the boy remained in perfect health. His wounds had healed and he was now well enough to go back to school. Alike to the parents of Joseph Meister and to the greatest medical men in France, this was a miracle.

Pasteur's name now became a household word in France, and in the remote parts of the world new hope was born. Rabies could be cured. Pasteur's development of the anti-rabies vaccine came as the climax of a great career. Rabies was not the first disease he had vanquished by vaccination, but it was his greatest triumph in the field of human disease. From all over France people who had been bitten by rabid dogs came to his laboratory; doctors in countries as far apart as Russia and America started sending him patients. The miracle was repeated over and over again, and nearly every single patient out of the hundreds who came for his treatment recovered perfectly. His 'hydrophobia service' became his main occupation, and until this had been organized on a big scale Pasteur's researches had to come to a full stop.

So there came into existence the Pasteur Institute in Paris. From all over France, from all over the world, subscriptions poured in towards the cost of setting up this great humanitarian centre; millionaires sent hundreds of pounds, and poor people sent their pennies. The subscription list from Alsace-Lorraine included the name of little Joseph Meister—the first human being to be brought back from the jaws of a rabid death. As Pasteur's friend Professor Vulpian said: "This new benefit adds to the number of those which our illustrious Pasteur has already rendered to humanity. Our works and our names will soon be buried under the rising tide of oblivion: the name and the works of Monsieur Pasteur, however, will continue to stand on heights too great to be reached by its sullen waves."

How true were those words of Professor Vulpian. The name of Pasteur is indeed immortal. In many cities there are Pasteur Institutes, and we also recall this great French scientist every time we use such terms as *pasteurization,* which is the method of food preservation which he perfected. His esteem among fellow scientists was beautifully expressed in this tribute from an English surgeon: "He was, it seems to me, the most perfect man who has ever entered the kingdom of Science. Here was a life, within the limits of humanity, well-nigh perfect. He worked incessantly; he lived to see his doctrines enthroned, his methods applied to a thousand affairs of manufacture and agriculture, his science put in practice by all doctors and surgeons, his name praised and blessed by mankind; and the very animals, if they could speak would say the same."

II

Although Pasteur's work revolutionized medicine, and opened up new realms of biological thought, he started his career neither as doctor nor biologist. His first researches belonged to chemistry. But he was a pioneer in every field he touched; indeed, when Fleming discovered penicillin it turned out that Pasteur had already discovered the first antibiotic, as long ago as 1877.

Louis Pasteur was born on December 27, 1822. His birthplace was Dôle in the eastern part of France. His father, Jean Joseph, had served as a conscript in Napoleon's army and had fought through the Peninsular War against Wellington's troops. He belonged to the Third Regiment of the Line—'the bravest among the brave', as one historian described it. Back in France his division fought with great courage against a force that outnumbered them five to one and Jean Pasteur, now a sergeant, was awarded the cross of the Legion of Honour. Came Waterloo and the cessation of hostilities, and Pasteur's father went back to his pre-war trade—that of a tanner. Always he kept his sabre as a reminder of the military glory of France's First Empire, and he never tired of recounting his army adventures to young Louis, who, as ardent a patriot as his father, was to add to the glory of France in no uncertain fashion.

There was nothing precocious about Louis; his boyhood was not exceptional and it gave no hint of the rich streak of genius which was to emerge later. The earliest talent which he showed was not for science at all, but for art; his pastel drawings were quite remarkable and he would certainly have made his mark as an artist. It was the headmaster of the college at Arbois who first realized there was a hidden spark in the conscientious and hardworking schoolboy. At that time his caution obscured his underlying brilliance. As a boy he never affirmed anything of which he was not absolutely sure—an invaluable trait for the future scientist, for it is a scientific axiom that' it is better to be sure than sorry'. (No scientist can afford to guess; he must always be quite certain.) Pasteur's headmaster inspired him with the ambition to become a professor, a prospect which made his father extremely proud. "If you can become a professor at Arbois, I shall be the happiest man on earth," he told young Louis. In 1840 he obtained his bachelor's degree and the examiners reported that he was 'very good in elementary science'. His target now was to enter the Ecole Normale Superieure in Paris, the great college that Napoleon had founded in 1808 for the training of young professors. The entrance examination was a stiff one and required months of intensive study; at that time Pasteur found the physics and chemistry which he had to learn most interesting, but he was not so happy about the mathematics, which he found both dry and exhausting. He entered the Ecole Normale in 1843. Paris was then the most important scientific centre in the world and Pasteur came under the influence of some of the great masters.

III

Chemistry became his passion and his first great discovery was made in that field. He had tackled a problem that had baffled scientists for years. This was concerned with the substance called tartaric acid, which was prepared commercially from tartar, the crust or sediment which accumulates in wine casks. Chemists had discovered that this acid exists in two quite different forms; the first was *tartaric acid* proper, and the other they called *racemic acid* or para-tartaric acid. These two acids had exactly the same composition; they contained exactly the same variety of atoms (namely carbon, hydrogen and oxygen) and these were present in identical proportions in both acids—namely four carbon atoms to six hydrogen atoms to six oxygen atoms, the formula being $C_4H_6O_6$. From the two acids, salts were easily prepared and crystals of these salts collected. The two sets of crystals—one set from tartaric acid and the other from racemic acid—looked as though they had precisely the same shape. Yet when the solutions of the one and of the other were tested in an instrument called a saccharimeter (which measures the amount of sugar in a solution by the extent to which the solution rotates a beam of polarized light) they behaved quite differently. The solution of tartrate twisted the light to the right, yet there was no twist at all when Pasteur repeated the experiment with racemic acid solution in his saccharimeter; to use the proper expression, the racemic was 'optically inactive'. Pasteur then proceeded to examine the crystals of the racemic more closely and found that they were not all of the same kind—some were right-handed crystals and the rest left-handed, the first class of crystal obviously being the mirror-image of the second class. (You can visualize this by thinking of the two different gloves that make up a pair.) So, patiently, by hand, Pasteur picked out all the right-handed crystals and put them in one pile and then he collected a pile containing nothing but left-handed crystals. When he dissolved the right-handed crystals he found that the solution twisted polarized light to the right; with the second set of crystals he obtained a rotation to the left. His 'right' tartrate performed in exactly the same way as the tartrate which he prepared from proper tartaric acid.

Next Pasteur put equal quantities of the 'right' and 'left', tartrate solutions together in his saccharimeter. This time there was no rotation of the polarized light; the twist which the 'right' tartrate would have produced was exactly balanced by the twist in the opposite direction imposed by the 'left' tartrate. In other words he had succeeded in re-combining the two optically active tartrates to produce the optically inactive racemate. Pasteur's discovery became the talk of the scientific world and won him the friendship of Monsieur Biot, one of the elder statesmen of French science. Largely through his friend's influence he received the red ribbon of the Legion of Honour in 1853, when he was only thirty.

In spite of his obvious talents for research, Pasteur was posted to the University of Strasburg as professor of chemistry. There he married Marie, the daughter of the university's rector, and of Madame Pasteur it has been said that "she was more than an incomparable companion; she was his best collaborator", acting as his scientific secretary as well as looking after their home and family.

IV

His next great triumph was the synthesis of racemic acid, which he made by transforming the compound called cinchonin tartrate at a high temperature. This won him a prize of 1500 francs awarded by the Pharmaceutical Society of Paris, and most of this money he used to equip his laboratory at Strasbourg. His interest in the asymmetry of tartrate crystals was now to lead him far away from conventional chemistry and into the field of microbiology, where he was to eclipse all his previous work and gain the nickname of '*the Christopher Columbus of microbes*'. The abrupt switch that happened to his career came about in this way. He happened to be studying what occurs when a solution of a tartrate like ammonium tartrate ferments. He was working with a species of blue-green mould called *Penicillium glaueum*—which is a cousin of the mould that produces penicillin—and he fed the fungus both ordinary ammonium tartrate and ammonium racemate to see what would happen. The mould thrived on ammonium racemate, but the result was quite unexpected; the left-handed tartrate appeared, so evidently what was happening was that the mould was fermenting the right-handed tartrate and leaving the other tartrate entirely alone! Pasteur recommended this as the best way of preparing left-handed tartrate.

By now he was developing vivid and exciting ideas about asymmetrical substances which were optically active, and the part that such substances played in the chemistry of living organisms. It emerged that many organisms can distinguish between left-handed and right-handed molecules; for example 'left' tartaric acid is twice as poisonous to guinea-pigs as the 'right' acid; the 'right' form of asparagine possesses a sweet taste whereas you cannot taste the 'left' form at all. Pasteur figured that asymmetrical compounds are inseparable from life and the researches of generations of organic chemists and biochemists have revealed how right he was in this idea.

V

His next university appointment was at Lille. This was the richest centre of industrial activity in Northern France, and the university placed special emphasis on training young men to become foremen and industrial managers. Pasteur entered a new and exciting world. He realized the great part which science could play in industry, and the inspired view which he showed to local manufacturers and to his students alike can be glimpsed in these remarks of Pasteur on applied science : "Where in your families will you find a young man whose curiosity and interest will not immediately be awakened when you put into his hands a potato, when with that potato he can produce sugar, with that sugar alcohol, with that alcohol ether and vinegar? What student would not be happy to tell his family in the evening that he has just been working on an electric telegraph ? And be convinced of this: such practical work is seldom if ever forgotten." It was men like Pasteur who started up the laboratories in colleges and schools that are so indispensable to the teaching of science. But while Pasteur spoke to industrialists of the practical value of science, he was at great pains to stress that *pure* research—knowledge for its own sake—is essential to the progress of science. "What is the use of a pure scientific discovery?" he was asked. He regarded the question as fatuous and quick as a flash he came back with the counter-question: "What use is a new-born baby?" That settled the argument.

Local industry quickly took to consulting Pasteur on their problems. There was, for example, the manufacturer who brewed alcohol for industrial purposes from sugar beet; initially he had obtained a good yield but for some unaccountable reason the process had gone wrong. Pasteur's first move was to examine the brew under a microscope, the instrument which symbolized his series of great studies on the phenomena of fermentation. Later it was he who introduced the microscope into English breweries! The microscope showed him that a good alcoholic brew contained nothing but round globules—the cells of the yeast plant. When the fermentation went awry, then along with the yeast the microscope revealed small rod-shaped bodies—the bacillus which converted sugar into lactic acid. As a chemist, Pasteur proceeded to dig far deeper into the chemistry of fermentation than anyone had done before. He soon discovered, for example, that ordinary alcoholic fermentation produces quite significant amounts of two chemicals—glycerine and succinic acid. The upshot of his researches in this field was that all the industries that used fermentation in making their products could now be re-organized on a scientific basis, a step which improved their output and the quality of their products. Not only did this effect brewing and bread-making (both of which involve the use of yeasts), but also such things as the manufacture of cheese and the curing of tobacco. Today all kinds of chemicals can be made by fermenting such materials as sugar and starch; such substances include acetone, citric acid, and the various antibodies (including penicillin).

VI

Pasteur established the principle that there is *'no fermentation without life'*. In other words, all fermentations are the result of the activity of microbes. As logical as Pasteur's move from tartaric acid crystals to ferments was his next switch—to the great biological question of what is called *'spontaneous generation'*. This had been a riddle for hundreds of years. A few wise men had reached the conclusion that only living things can beget living things; but the majority believed the opposite—that living organisms—plants and animals—could suddenly, spontaneously, appear. For example one first-class scientist, van Helmont, maintained that he could produce mice by putting some cheese on a pile of dirty linen and leaving the mysterious force of spontaneous generation to conjure forth baby mice. Another classic delusion was the belief that driftwood spontaneously produced barnacles (the kind called 'goose barnacles') and that these changed spontaneously into butterflies, which in turn were transformed into birds.

Once in a long while someone would do a scientific experiment that clearly contradicted this faith in spontaneous generation. There was the Italian naturalist Francesco Redi who did not accept the current view that meat spontaneously generated maggots. They were, he said, nothing but the larvae of flies. To prove his case he protected some meat with a piece of gauze that excluded the flies; so long as the flies were unable to get at the meat and lay their eggs on it, the meat never turned maggoty. (This was, of course, the origin of the still familiar meat safe.) Men like Redi disproved the theory of spontaneous generation and substituted the rule that *'like begets like'*—in other words, a mouse can only be derived from other mice and cannot come into existence in any other way.

But when the microscope was invented and a new world- the world of microbes—was discovered a new race of believers in spontaneous generation came on the scene. They honestly believed that the apparently simple microbes can suddenly materialize in such liquids as broth, milk and sugar solution. Pasteur realized that there would be no progress in micro-biology unless this question was settled once and for all. The spontaneous generationists believed that life could be conjured up out of thin air. Pasteur pondered upon this idea and came to an entirely different conclusion. Suppose that the air was not thin at all; suppose the atmosphere contained microbes or the seeds *(spores)* of microbes. Then if you put down a saucer or bottle of milk, surely some of those microbes or their spores will soon get in the milk and proceed to multiply until the milk is stiff with them. So Pasteur first boiled some milk in a flask to kill any microbes already in it. Then he plugged the neck of the flask with a plug of cotton-wool which had been singed in a flame to destroy any microbes in it. This simple device incidentally would allow air to reach the milk but that air would be filtered, any microbes being trapped by the cotton-wool. What was the result? Instead of going bad as milk normally would, it remained fresh indefinitely; it did not ferment and no microbes appeared in it.

In London's Royal Institution a British physicist, John Tyndall, did similar experiments and the results he obtained fitted in perfectly with Pasteur's. These two men did hundreds of tests and always obtained the same results, but still the theory of spontaneous generation survived. Their cleverest opponent, a Frenchman called Pouchet, had produced one very awkward fact on which the spontaneous generationists made their last great stand. Pouchet had brewed an infusion of hay but when he sterilized this liquor according to Pasteur's method and plugged the flask with a sterile filter the hay liquor proceeded to go bad! One fact like that can destroy a scientific theory, as both, Pasteur and Tyndall knew all too well. It was Tyndall who found the correct explanation. He repeated Pouchet's experiment; even after boiling the hay infusion eight solid hours, Tyndall could still not get it to keep. One up to Pouchet, but Tyndall struck back with a new series experiments which gave the final death-blow to spontaneous generation. He proved that the heating killed only the living hay bacteria. Their *spores* on the other hand could survive being boiled! To kill the whole lot it was necessary to boil once, allow the liquor to cool, let the spores germinate and produce living bacteria, and then boil a second time. The second boiling did the trick and proved effective even if it last, for only a fraction of a second. (This method of 'discontinuous heating' is known as *tyndallization* and is complementary to *pasteurization*,)

These techniques of sterilization were soon applied to industry. Pasteur perfected a method of preserving wine, beer and other liquors that can go bad by heating them for a few minutes at 50 degree Centigrade or so. This method—now universally known as pasteurization—is used all over the world; most milk in our country nowadays is pasteurized.

VII

No scientist has ever possessed a more wide-ranging brain than Pasteur's. After he had realized that microbes bring about the decay of dead tissues—butcher's meat, for instance – he started thinking about what microbes might do to the bodies of living animals. Particular kinds of fermentation were due to particular kinds of microbes; was it possible that the contagious diseases of man and other animals were due to specific microorganisms? This was Pasteur's germ theory of disease which was to revolutionize medicine.

The first infectious disease he tackled was *pébrine*. This is a disease of silkworms that was ruining the French silk industry. Pasteur discovered that this was due to a microbe (called *Nosema bombycis*). He showed also that the disease was propagated by healthy worms eating with their food the frass of infected caterpillars; the moths from infected caterpillars laid infected eggs. He then taught the silkworm cultivators how to obtain uninfected eggs and so he saved this industry.

The time was ripe for Pasteur's incursion into medicine. The introduction of anaesthetics had made it possible for surgeons to perform complicated and time-consuming operations successfully. But too many patients died after the operations because little could be done to protect them against blood poisoning. As one surgeon said of the surgical wards of that period, "Pus seemed to germinate everywhere, as if it had been sown by the surgeon." There was almost complete ignorance about the mechanism whereby a cut became infected and proceeded to suppurate. A few daring surgeons tried sterilizing the flesh with such things as hot brandy and the bandages they used with an antiseptic mixture of alcohol and water. The results were most encouraging and one of these surgeons, Alphonse Guerin, teamed up with Pasteur in 1873 with the aim of improving his antiseptic methods. Pasteur had another vigorous disciple in Britain—Joseph Lister (1827-1912), who worked out his principles of antiseptic surgery in Glasgow and Edinburgh. Before the onslaught of these men '*the evil of putrefaction*' as Lister called it practically disappeared from the surgical wards and the mortality after operations dropped to a tenth of what it had been. The French Government now recognized the great achievements of Pasteur by giving him a life pension.

German scientists, too, had become imbued with Pasteur's ideas about germs and one of them — Robert Koch (1843-910) — gave the new science of medical bacteriology a flying start by discovering the full details of the life history of the anthrax bacillus.

He followed that up by two other vitally important discoveries—the isolation of the bacteria that cause tuberculosis and cholera. Then followed the golden era of medical bacteriology; inside the next twenty years the things that cause most bacterial diseases were identified and studied.

VIII

The next problem which Pasteur tackled was the question of immunity to disease. Could the immunity of human beings to attack by specific germs be produced by any artificial means? Edward Jenner had immunized people against smallpox by vaccinating them with the mild form of the disease known as cowpox; that was back in 1798, and no other useful vaccine had been discovered since.

The first vaccine which Pasteur produced was the result of a lucky accident, the kind of accident which the brilliantly prepared minds of master scientists exploit to the full. In 1879 an epidemic disease was crippling poultry keeping in France. Pasteur turned his attention to this disease. He showed that the germ of this chicken cholera could be grown on broth prepared from chicken gristle. By accident a culture of this microbe on broth was lost. Some weeks later it came to light again and Pasteur carried out a routine test on it, injecting it into some hens to see whether it was still infectious. The hens fell sick as usual—but then, to everyone's delighted surprise, they began to recover. The stale culture was an effective vaccine. But how could he produce the vaccine again when he wanted it ? Pasteur discovered that the strength of a cholera culture could be reduced by starving the bacteria of air; the weakened *culture — attentuated culture* is the technical term for this — protected the hens which Pasteur vaccinated with it.

The most serious disease of farm animals at that time was anthrax—or *charbon* as the French called it. He found he could grow the *germ—Bacillus anthracis*—in sterile urine and broth. Pasteur's next problem was to attenuate the bacillus to produce a safe and effective vaccine. He knew that at 45 degrees Centigrade this bacillus cannot grow at all; it should therefore be possible to pick a somewhat lower temperature that would enable the bacillus to multiply but provide a culture rather weaker than normal. Attentuation by heat proved successful and Pasteur achieved the much-needed vaccine. In May 1881 he demonstrated the power of his anthrax vaccination. An agricultural society put at his disposal a farm at Pouilly le Fort and 60 sheep. Pasteur vaccinated 25 sheep, keeping another batch of 25 animals unvaccinated for the sake of comparison. Then he injected all the sheep with virulent anthrax. Almost all of the unvaccinated sheep died but, exactly as Pasteur predicted, every single vaccinated sheep survived. "*No success had ever been greater than Pasteur's,*" wrote his biographer. "*The veterinary surgeons, until then the most sceptical, now convinced, desired to become the apostles of his doctrine (of vaccination).*" The vaccine caught on like wildfire and in a matter of a fortnight Pasteur and his assistants, Roux and Chamberland, treated no less than 20,000 sheep, with complete success. Anthrax was conquered, one of the four animal diseases for which Pasteur perfected vaccines in the incredibly short period of four years.

    The scientist had been working on rabies which medical men had always regarded as incurable. By 1885 he had been able to produce a vaccine that gave protection to dogs. Pasteur worried about making the final test that was necessary—the trial of his vaccine on a human being. The effects of this disease in humans were horrifying in the extreme and Pasteur could not forget the case of rabies he had once seen in hospital. He hesitated to take the plunge; as he wrote on March 28, 1885, " I have not yet dared to treat human beings, but the time is not far off, and I am much inclined to begin by myself—inoculating myself with rabies and then arresting the consequences. "Pasteur had never lacked courage and had never hesitated to come in contact with the most dangerous of the contagious diseases; now he was prepared to take the ultimate risk. But it never became necessary, because of the arrival of little Joseph Meister, the lad from Alsace, at his laboratory. The test as you know was made on the boy and saved his life. The story of Pasteur's life, which began in 1822 and ended in 1895 is now complete. No scientist ever achieved more in a lifetime and I hope you will want to read more about Pasteur ; if so, there is a wonderful biography about him, *The Life of Pasteur* by René Vallery-Radot, his son-in-law.

Meghnad Saha

(1893–1956)

A Great Name in Astrophysics

Meghnad Saha was born in a village of Dacca district, now in Bangladesh, on October 6, 1893. His birth took place in the midst of a thunderstorm, with furious rains lashing at the hut where the family lived. So he was named. Meghnad which means 'the sound of thunder'.

His father was a petty grocer who barely managed to keep his large family from starvation. He wanted his fifth child Meghnad to start earning for the family from child-hood. So when the boy completed primary education, he kept him at the grocery shop as a helping hand. But Meghnad was a brilliant boy and his teachers and the medical doctor persuaded the grocer to send him to boarding school eleven kilometers away, the doctor offering to pay his school fees.

The young Meghnad received a scholarship when he stood first in the middle school examination. This enabled him to go to Dacca to join the high school. But he, along with some other boys were expelled from the school for boycotting the Governor's visit to the school. It was a protest against the partition of Bengal and in boycotting the visit the students were joining the popular agitation against the British.

Saha suffered more than the other boys for he was not expelled but lost his scholarship as well. He joined another school,

Kishorilal Jubilee School. The school authorities not only gave admission to him, but knowing his family's poverty, secured for him another scholarship.

This kindness of the school management was in due course rewarded when the boy brought a good reputation to the school by getting a first rank in the final examination in the entire East Bengal. This won him, again, a scholarship. He went to Calcutta and joined the Presidency College there, the most prestigious institution for higher learning in Bengal in those days.

This was a remarkable period of his life which moulded his thinking, character and personality Among his professors were the reputed scientists J.C. Bose and P.C. Ray. Subhash Chandra Bose was his hostel mate, and so were P.C. Mahalanobis and S.N. Bose who like him became noted scientists.

Saha could get only the second rank in the B.Sc. and M.Sc. examinations. In M.Sc. the first rank went to his friend S.N. Bose.

He wanted to join government service to help his needy family. But he found that every government post was barred to him because, as a student, he had boycotted the Governor's visit and had associated with active nationalists like Subash Chandra Bose and Rajendra Prasad.

His family had placed much hope in him. His mother had sold her gold bangles to pay for his examination fees. Now his younger brother also was staying with him in Calcutta and he realized that he should at least earn enough money to maintain himself and his brother. He managed to do this by giving tuitions for a time. This was excruciatingly hard work walking the length and breadth of Calcutta and cycling miles and miles in the big city.

Fortunately, a great man, Ashuthosh Mukherjee was the Vice-Chancellor of Calcutta university at that time. Dr. Mukherjee had already known them as brilliant students. He appointed both Saha and S.N. Bose as lecturers in the newly started University College of Science.

But the college provided no facility for research and what Saha and Bose wanted most was to conduct research. As a consolation, they decided to jointly translate Einstein's *'Theory of Relativity'* from German into English. As it happened, this was the first English translation of the theory in the entire world !

Thermodynamics, relativity and atomic theory were new developments in physics at that time. Once, while preparing his lecture notes, Saha came across a problem. The problem was about the origin of the spectral lines. Spectral lines are the dark and bright lines of varying thickness found in the spectrum produced when sunlight passes through a glass prism.

Gustav Robert Kirchhoff, the eminent scientist, had shown that these spectral lines tell us about the elements that compose the sun. While it was known that the spectral lines represent the elements of the sun, it remained unknown how the lines came to be present in the spectrum.

Saha presented his 'ionisation formula' which explained the presence of the spectral lines. The formula also enabled astronomers to know the temperature, pressure and other aspects of the sun or the interior of any star. The formula was a breakthrough in astrophysics.

He was only twenty-five at the time. He became a world famous scientist with this discovery. An eminent astronomer called it the twelfth major discovery in astrophysics.

In 1940, Otto Hahn discovered that the nucleus of the atom can be split. Saha immediately recognized its importance. On his initiative Nuclear Physics was made a subject of study, for the first time in India, at Calcutta University. And in 1948, he founded what is today known as the Saha Institute of Nuclear Physics in Calcutta.

He fought and won an election to the Parliament against the candidate of the Indian National Congress, and as a member of the opposition, he raised his voice of protest against lapses of the government, particularly in the areas of education, science, atomic energy, and industrialisation and asserted himself to uphold the democratic right of the people. Saha was elected to the Royal Society of London, an honour coveted by scientists the world over, in 1927.

Seven children — three sons and four daughters — were born to him and his wife Radhamoni Roy. His eldest son Ajit Saha became a physicist like him. Meghnad Saha passed away on February 16, 1956.

Michael Faraday

(1791–1867)

Remarkable Genius

Electricity has ceased to be an intangible phenomenon, it has become a domestic and industrial slave. Michael Faraday made this possible. His discovery of induced currents, which meant the production of electric current from magnetic force, was the dawn of a new epoch. By it he made the way clear for his followers to produce the dynamo, the generator of electric current: indirectly, we owe to him electric light and power, the telephone, telegraphy, and a thousand other devices made possible by tapping the vast sources of electricity which he discovered. On his work were based the researches which led to wireless telegraphy. His experiments in electrolysis were invaluable: he magnetized light: he was a great chemist as well as physicist. A humble, patient, brilliant scientist, his achievements are among the highest ever attained by humanity, for the fruits of them may never be exhausted: "The progress of future research will tend, not to dim nor to diminish, but to enhance and glorify the labours of this mighty investigator."

Sir Humphry Davy, the great chemist, was almost at the end of his days, when a friend, as the story goes, asked him to declare which of his discoveries he considered to be the greatest. Davy, who was somewhat jealous of his fame and professional reputation, considered for a while. Then he gave a short list, named an electrical discovery, the results of some of his chemical researches, his famous safety lamp, and concluded, "But the greatest of all my discoveries was the discovery of Michael Faraday." And Davy, who died in 1829, spoke more truly than he knew.

Michael Faraday was born at Newington Butts on September 22, 1791. His father, James Faraday, was a blacksmith who had come to London from Yorkshire. He was hard put to it to make end meet, and the young Michael experienced the hardships of poverty throughout his childhood.

Some five years after Michael's birth, the family moved to Jacob's Well Mews, Charles Street, Manchester Square. From there he was sent to school. His education, he said years later, was of "the most ordinary description, consisting of little more than the rudiments of reading, writing and arithmetic at a common day school. My hours out of school were passed at home and in the streets." When he had become famous, Faraday would often frequent that part of London in which his early days had been spent, and point out the places where he had looked after his younger sister, or had played marbles in the street. But the young Michael was not allowed to run wild. His father had become a member of the Sandemanian Sect, and Faraday, as his later life showed, absorbed his deep religious convictions.

At the age of thirteen, Michael Faraday had to leave school and go out to earn his living. He became an errand boy for neighbouring book-seller, named Riebau. His job consisted mainly of carrying and delivering newspapers; in the words of Professor Tyndall, Faraday's friend and successor, "he slid along the London pavements, a bright-eyed errand boy, with a load of brown curls up his head and a packet of newspapers under his arm."

His trial year as an errand boy was so satisfactory that, in October 1805, Faraday was apprenticed, without premium, to Riebau to learn the trade of bookbinder and stationer. For the next few years Faraday was busy, for, in addition to learning his trade, he had access to books, and Riebau, who was a considerate master, allowed him occasional leisure to attend shilling lectures on natural philosophy. Faraday made friends at the lectures and learned quickly from the books and papers: his interests turned particularly to science and chemistry. He was soon inspired with the ambition to leave trade and take up a scientific career.

And then occurred an incident which was the turning point of his life. The story has often been dramatized in the telling; hear it the simplicity of Faraday's own words : -

"During my apprenticeship I had the good fortune, through the kindness of Mr. Dance, who was a customer of my master's shop, and also a member of the Royal Institution, to hear four of the last lectures of Sir H. Davy in that locality. The dates of these lectures were February 29, March 14, April 8, and April 10, 1812. Of these I made notes, and then wrote out the lectures in a fuller form, interspersing them with such drawings as I could make. The desire to be engaged in scientific occupation, even though the lowest kind, induced me, whilst an apprentice, to write, in my ignorance of the world and simplicity of my mind, to Sir Joseph Banks, then President of the Royal Society. Naturally enough, 'no answer' was the reply left with the porter."

Faraday sent the notes of Davy's lectures, which he had absorbed with interest and delight, to Sir Humphry himself, and asked to be allowed to quit trade, which he thought "vicious and selfish," and to devote himself to science. Davy replied courteously.

One evening, when Michael Faraday was undressing to go to bed, there was a loud knock on the door. Davy's carriage was outside, and Davy's servant handed Faraday a note, as a result of which he called next morning at the Royal Institution, where he was engaged as a laboratory assistant at twenty-five shillings a week. Faraday at once became a valuable help to Davy, and the two experimented together with chlorine. During his work at this period both he and Davy were several times injured by explosions of chloride of nitrogen, a dangerous gas.

In the autumn of 1813 Sir Humphry Davy and his wife went abroad, and Faraday accompanied them, partly as secretary and assistant, partly as valet. Davy, who was Faraday's life-long friend, treated his assistant with every consideration, but Lady Davy used him as a menial, and almost drove him to return home. At Geneva, Davy was the guest of his friend, De La Rive, father of the great electrician, who was shocked when he found out that Faraday, then living with the servants, was actually Davy's laboratory assistant. He at once suggested that Faraday should eat with the family, but Lady Davy objected; so the courteous host compromised by sending Faraday's meals to his own room.

Faraday was away two years with Davy. He kept an interesting journal of their tour through France, Switzerland, Italy and the Tyrol, and his affectionate letters to his mother at this period are typical of the man. One of the most interesting notes of the tour records how on June 17, 1814, he saw M. Volta, who came to Sir H. Davy, a hale, elderly man, bearing the red ribbon, and very free in conversation.

The travellers returned to London in 1815 and Faraday was re-engaged at the Royal Institution. He now settled down to a life of research and discovery: " From May 7, 1815 (the date on which he rejoined the institution), onwards, his life was a time of steady intellectual growth, spent in chemical research, in the explaining of phenomena, and in what is by no means his least claim on our regard, the popularization of scientific knowledge."

In June, 1821, Faraday married a Miss Sarah Barnard, the daughter of an elder of the Sandemanian Church. The union was long and happy. "A month after his marriage," writes Bence Jones, in his *Life and Letters of Faraday,* " he made his confession of sin and profession of faith before the Sandemanian Church. When his wife asked him why he had not told her what he was about to do, he only replied, 'That *is* between me and my God. '

Faraday's chemical researches were first made in directions opened up by Davy, and soon resulted in valuable advances and discoveries. He made a particular study of chlorine. On one occasion Dr. Paris entered the laboratory while Faraday was at work and, seeing an oily liquid in a tube, censured the scientist for his carelessness in using dirty vessels. The next day Dr. Paris received a note :—

"Dear Sir,

The oil you noticed yesterday turns out to be liquid chlorine.

Yours faithfully,

M. Faraday"

Faraday had succeed in converting chlorine gas into a liquid by means of its own pressure. This was an important discovery, and led to many experiments with other gases, which produced like results. During this period of his life, Faraday, besides liquefying several gases, made the first rough experiments on the diffusion of gases, a phenomenon first pointed out by John Dalton. He discovered two new chlorines of carbon; investigated alloys of steel; produced several new kinds of optical glass; and in 1825 announced the discovery of benzol. In addition to these valuable discoveries, there must be credited to the careful, accurate and painstaking Faraday the general improvement in laboratory methods which dates from this time.

In 1823 Faraday was elected a Fellow of the Royal Society, and two years later was made director of the laboratory at the Royal Institution. In 1833 he was appointed Fullerian Professor of Chemistry in the Royal Institution for life, without the obligation to give lectures.

Meanwhile he had been conducting experiments and making the first of those electrical discoveries which have made his name a household word. Basing his experiments on the work of Oersted and Wallaston in the realm of electromagnetism, Faraday explained the continual rotation of a magnet and the electrified wire round each other. That was in 1821. Ten years later he made his great discovery of electrical induction; "Faraday in 1831," writes

Tyndal, "had now reached the threshold of a career pf discovery unparalleled in the history of pure experimental science."

In November, 1831, Faraday announced his epoch-making discovery to the Royal Society.

He showed that an electromotive force is setup in a conducting wire when it is moved at right angles to a magnetic field. If the wire is part of a closed circuit, its motion results in an "induced" current.

Now, in 1824, Arago had discovered that a disk of non-magnetic metal had the power of bringing a vibrating magnetic needle suspended over it to rest, and that on causing the disk to rotate the magnetic needle rotated with it. When both were quiescent, there was no measurable attraction or repulsion exerted between the needle and the disk, yet when in motion the disk was competent to drag after it, not only a light needle, but a heavy magnet. To this, "the darkest physical phenomenon of that day," Faraday turned his attention. He placed a copper disk between the poles of a large horse-shoe magnet and, connecting the axis and the edge of the disk each by a wire with a galvanometer (an electric current-measuring machine), he obtained, when the disk was turned round, a constant flow of electricity.

Faraday then stated the law governing the production of currents in disks and wires. When iron filings are scattered over a magnet, the particles of iron arrange themselves in certain definite lines called magnetic curves. Faraday named these curves "lines of magnetic force"; and he showed that to produce induced currents neither approach to nor withdrawal from a magnetic source, or centre, or pole was essential, but that it was only necessary to cut appropriately the lines of magnetic force.

Or, to sum up the essence of electro-magnetic induction: electric current is induced in a conductor which cuts the lines of magnetic force, or whenever the number of these lines passing through the current of the conductor is in any way varied: any movement which causes an alteration to take place—in the amount of magnetic conduction through the coil produces a transient current, the electromotive force of which is proportional to the rate of the alteration.

In fact, Faraday had tapped new and inexhaustible sources of electricity. Before these experiments, magnetism had been evoked by electricity; he had aimed, and succeeded, at exciting electricity by magnetism.

Faraday's spinning disk between the poles of the magnet— his "magneto-electric machine"— was, in fact, a primitive dynamo: he was the father of commercially practical generation of electricity.

Through this discovery of induced currents was made possible the invention of the telephone, the development of the telegraph, of electric lighting, and the production of electricity for the thousand and one uses of modern life.

Medicine was quick to benefit by the discovery, and faradic electricity is extensively used in medical practice. Faradic electricity is produced by a simple form of induction coil, consisting of a primary coil of thick wire bound round a tube in which slides a bundle of iron wire. A secondary coil of fine wire is fitted over this. The induced or faradic current is generated in the secondary coil only when the primary current makes or breaks contact (as Faraday discovered in his experiments), so an automatic "interrupter" is used to make and break the primary current and produce a flow of induced current. Faradic electricity was used in the diagnosis of nervous and muscular disorders, for the treatment of inflamed and painful joints in acute cases of articular rheumatism, and to combat neurasthenia and functional disorders.

Faraday himself was partially responsible for an early adoption of electric light. In 1858 Professor F. H. Holmes built an improved and powerful electro-magnetic machine on the pattern invented by Faraday, and Faraday, who in 1836 had been appointed scientific adviser to Trinity House, suggested the use of it for generating current for lighthouse lamps. Accordingly, a complete installation was made in the South Foreland Lighthouse, which was the first in the world to be lit by electricity.

But Faraday was not concerned in the commercial development of his discoveries—he even discarded the supplementary commercial work which added to his income. He went on with his experiments in the spirit of the true investigator. Four years later, investigating from observations made by a William Jenkin, Faraday discovered " extra current " or the current which is induced in the primary wire itself at the moment of making and breaking contact. In writing of this, Professor Tyndall made an illuminating comment on Faraday's thoroughness:—

"Seven and thirty years have passed since the discovery of magneto-electricity; but, if we except the extra current, until quite recently nothing of moment was added to the subject. Faraday entertained the opinion that the discoverer of a great law or principle had a right to the 'spoils' —this was his term—arising from its illustration; and guided by the principle he had discovered, his wonderful mind, aided by his ten fingers, overran in a single autumn this vast domain, and hardly left behind him the shred of a fact to be gathered by his successors."

In 1833 Faraday decided that the various kinds of electricity which had been discovered were all identical. Next he turned to electro-chemistry and electro-chemical decomposition, which he named electrolysis. At this time he invented a number of new terms : he abandoned the word "poles" for the ends of a decomposing cell and called them "electrodes" substances which could be decomposed by an electric current he called "electrolytes"; and constituents of the decomposed electrolyte became "ions". He was the first to make accurate quantitative experiments in electrolysis. The two great laws he formulated are: the mass of substance liberated from an electrolyte by the passage though it of electric current is proportional (1) to the total quantity of electricity which passes through the electrolyte and (2) to the chemical equivalent weight of the substance liberated. How valuable these discoveries have been may be imagined, when one considers the vast commercial uses of electro-chemistry in the refining of metals, metallic production from ore, and in the making of caustic soda, chlorates and gases.

It had been the intention of Sir Robert Peel to offer Faraday a pension, but a change in government left the task to Lord Melbourne, who performed it somewhat brusquely. Faraday proudly refused the offer until Melbourne sent him a written apology, and eventually he was granted £300 a year for the rest of his life.

In 1841 Faraday had a breakdown in health. His nerves were shattered and his mind was seriously shaken. For three years he did nothing, not even "reading on science." He visited Switzerland with his wife and brother-in-law. One of the entries in his journal gives us an insight on his warm affection for his family. At Interlaken, on August 2, 1841, he wrote: "Cloutnail-making goes on here rather considerably, and is a very neat and pretty operation to observe. I love a smith's shop and anything relating to smithery. My father was a smith."

There is a similar story told of how, when sitting for his bust to Noble, he was agitated when the sculptor made a clattering with his chisels. Noble apologized and said that no doubt the jingling of the tools had distressed him and that he was tired. " No, my dear Mr. Noble," answered Faraday gently, " but the noise reminded me of my father's anvil and took me back to my boyhood."

Faraday got back to work as soon as his health permitted, and in 1845 he announced his great discovery of the magnetic rotation of the plane of polarized light. He was the first to seek for and discover effects of magnetism on light; and he found that if polarized light is passed through a transparent substance placed in a strong magnetic field, the plane of polarization is rotated. This, in the science of magneto-optics, is known as the Faraday effect.

Faraday suggested that light might be a series of transverse vibrations travelling along his lines of electric and magnetic force, thus brilliantly foreshadowing the electro-magnetic theory of light later developed by Clerk Maxwell.

The discoveries of Faraday led to the work of Maxwell, and later of Hertz, in electro-magnetic radiation, which paved the way for the invention of wireless telegraphy.

All the experiments and discoveries of Michael Faraday, "the greatest experimental philosopher the world has ever seen," would take a volume to mention. Much of his later work was on diamagnetism: that is, he discovered that every substance reacts to a magnet; those that are attracted, like iron, he called paramagnetic; those that are faintly repelled, diamagnetic.

In all Faraday's work there appeared the same minute care in experimentation, the same wide grasp of possible theory and possible counter-theory, the same imagination, the intuitive grasp of developments, and finally the beautiful lucidity of his exposition. He smells the truth, "said Kohlrausch, but Faraday still tested the smell until it was inescapable. His fame as a scientist was equalled by his power and reputation as a lecturer. He exercised a magic upon his hearers, but he was at his best at the juvenile lectures which he delivered in the Christmas holidays.

For forty years Faraday lived in the Royal Institution, but in 1858 a house on the green at Hampton Court was placed at his disposal by the Queen at the instigation of Prince Albert. Here, at length, he retired, giving up his last active work in 1865. From then until his death on August 25, 1867, he spent his time, with failing memory, *"Just waiting."*

Michael Faraday's achievements, his discoveries which mark a tremendous stride in the progress of the world, have overshadowed the man. Short, curly-headed, active, vivacious, with a hearty laugh, he was a man of strong emotions. He was generous, charitable and deeply religious. His affection for his parents and for his wife was strong and lasting. For some time he was an elder in the Sandemanian Church, in which he was an earnest preacher. But those who knew him stress, not his intellectual power, not his sincerity and his straightforwardness, but his natural refinement and delicacy of character.

"Faraday himself," said Sir Ambrose Fleming, "never gave attention to so-called 'useful' applications of his scientific work. His mind was entirely occupied with the endeavour to penetrate further into the secrets of Nature. ...For nearly forty years he went every working day into his laboratory with some new question to put experimentally to Nature and he never paused until he had a sufficient answer, ' yea' or 'nay,' to his query."

There could be no finer summary of Faraday, the man, than the description of him given by one of his intimate friends: " Nature, not education, made Faraday strong and refined. A favourite experiment of his own was representative of himself. He loved to show that water, in crystallizing, excluded all foreign ingredients, however intimately they might be mixed with it. Out of acids, alkalis, or saline solutions, the crystal came sweet and pure. By some such natural process in the formation of this man, beauty and nobleness coalesced, to the exclusion of everything vulgar and low."

Pierre and Marie Curie

(1859–1906)
(1867–1934)

Discoverers of Radium

"Now, Bronya, I ask you for a definite answer. Decide if you can really take me in at your house, for I can come now. I have enough to pay all my expenses. if, therefore, without depriving yourself of a great deal, you could give me food—write to me and say so. It would be a great happiness, as that would restore me spiritually after the cruel trials I have been through this summer. Since you are expecting a child, perhaps I might be of use to you . . .

So wrote Manya Sklodovska from Poland to her elder sister Bronya, married and living in Paris. The sisters were children of a Warsaw professor whose refusal to teach in the Russian—the oppressor's—language had resulted in his dismissal. The family was a proud one, and happy. Their mother and the oldest sister had died when Manya was a small girl; those remaining had taken jobs as soon as they were old enough to sustain themselves and their father. Manya, the youngest became a children's governess at eighteen, working for a wealthy country family, landowners with the address, strange to English eyes and ears, of "Szczuki, near Przasnysz" : Manya's daughter, who, many years later, wrote her mother's biography, forbore to give their name. Here Manya managed to save a small sum each month and send it to Bronya, studying medicine in Paris.

When Bronya married a young, exiled Pole, she no longer needed Manya's help. Manya, baby of the family, decided she too must go to Paris and study. From childhood she had been fascinated by science, partly because it was one of the "higher studies" the Russians forbade Poles to engage in, one that people studied in cellars: now she was anxious to go to the only place in Europe where the subject was taught properly, the Sorbonne. Her decision was hastened by an unhappy love affair. A month after joining the family at Szczuki she had fallen in love with the son. The parents had seemed devoted to her, had treated her as an equal—which in all but money she was—but when young Casimir asked their permission to marry, they were horrified. As Manya's daughter says, in her biography:

"The fact that the girl was of good family, that she was cultivated, brilliant and of irreproachable reputation, the fact that her father was honourably known in Warsaw—none of this counted against six implacable little words; one does not marry a governess."

Bronya was overjoyed at the prospect of being joined by her "little sister". She and her husband (also a Casimir, Dr Casimir Dluski) were not rich, but if Manya would bring bedclothes, towels, mattress, stout shoes and both her hats, they would happily look after her; she would be warm, well-fed, and loved—and this, at the moment, was what Manya wanted as much as anything else in the world, affection and a chance to get down to study, to make something of herself.

She was not disappointed; the Dluskis treated her like a daughter, fussed over her, showed her how to enrol at the Sorbonne, entertained her. The house was never still, she was never lonely. Dluski would play the piano far into the night. Sometimes he would be joined by a friend, a young Polish pianist with a technique far superior to his own, whose name was Ignace Paderewski and who one day would become not only a world-famous musician but prime minister of a now, free, Poland.

Manya flung herself into the science course at the Sorbonne, though at first she was too shy to mix with the French students, would talk only to the Poles. She was disappointed in her command of the French language; she had hoped it was up to the demands of this new study, but she found herself misunderstanding or failing to comprehend whole sentences in the lectures. She found that, apart from homework in science, she had to devote precious time to mastering the language in which it was taught, and now, sadly, she realized that she would have to leave Bronya's household: the social life was too much, she could not take part in it and have any hope of passing her exams. She was constantly being interrupted by music, by well-intentioned conversation, by the sound of patients bursting into the surgery at all hours. Moreover she never heard a word of French.

The Dluskis were upset, not a little offended, but they arranged for her to move to lodgings, gave her presents to take with her, to brighten up the bleak room she had chosen, kissed her farewell.

This move, which in many ways had been as big a step as the journey from Poland, marked a step in the young girl's life. From now on she was alone, without any sort of human entanglement; she would devote her life to study: furthermore, she would now be "Marie" Sklodovska, writing her name in the French style. It was simpler, would avoid foolish questions: Later, there would be time for nationalism; now all that mattered was study.

Marie had set her heart on two Masters' degrees, in physics and in mathematics, and in a very short time she had them. Shortly afterwards she was awarded a scholarship which would let her stay a little longer in Paris and she was overjoyed. She paid a quick visit to her father in Warsaw, then returned, intending to find research work which would provide the money to lengthen her stay indefinitely.

Shortly afterwards, the Society for Encouragement of National Industry ordered a study from her on the magnetic properties of various steels. She was delighted at being offered the work, but it required cumbersome equipment, far larger then she could fit into her own small laboratory. She was told to contact a young French scientist, Pierre Curie, who was doing very advanced work in his well-equipped laboratory: he, perhaps, might be able to find room for her. She was shy of this approach, and so a friend asked them both to tea. As their daughter, Eve, was to write, many years later:

"How strange it was, the physicist thought, to talk to a woman of the work one love, using technical terms, complicated formulae, and to see that woman, charming and young, become animated, understand . . . "

Marie was twenty-seven, Pierre thirty-five. He had been born in Paris, 15 May, 1859, the son of a doctor. He had never been to school because the father, deciding his mind was too unorthodox for school life, had educated him at home with the aid of a remarkable tutor. At sixteen the boy had become a Bachelor of Science. Three years later he was appointed Laboratory Assistant at the Sorbonne. With his brother Jacques he announced the discovery of Piezo-electricity, which was to open up new fields in the accurate measurement of electricity and, later, in its use for the transmission of sounds. This was the man who fell in love with Marie Sklodovska—but her reply disappointed him. No, she would never marry a Frenchman, it would mean leaving her family for ever, abandoning Poland; it was unthinkable. But, says their daughter, "Pierre Curie gradually made a human being out of the young hermit." She changed her mind and in 1895 they were married.

In 1897 they had their first daughter, Irene, a beautiful child who, like her parents, was destined one day to be a winner of the Nobel Prize.

For some time the scientist Henri Becquerel had been experimenting with the element uranium. He had discovered that a compound of it placed on a photographic plate made an impression, even through a sheet of black paper. He decided that the element emitted a form of radiation, not unlike X-rays, which could penetrate where light would not go. Both Pierre and Marie Curie were fascinated by this phenomenon and Marie, using equipment invented by her husband and his brother, set out to measure its extent. Step by step she catalogued the peculiarities of the radiation, decided it must be some "atomic" property, found that it could be observed in compounds of another element, thorium. Because this peculiar property for convenience had to have its own special name she christened it, "radioactivity". In search of other substances which might have the same property, she selected, with Pierre, samples from the mineral collection at the School of Physics. Here came a dramatic discovery. When she found radioactivity it was a great deal stronger than with either uranium or thorium. It must therefore be some other element—but Marie Curie had examined every one known.

The answer was clear to her: she was on the track of a new element. Pierre decided to interrupt his own research to help her isolate it. In the spring of 1898 there began a collaboration which lasted eight years. They began by investigating an ore of uranium, pitch-blende, even though its composition had been known for years. They reasoned that the new element would be in it; they calculated, with what they hoped was pessimism, that it would be present to a maximum quantity of one per cent. (In fact it was a millionth part.) Within a month of the start of their collaboration they were able to isolate a radioactive element, which in honour of Marie's country they called "polonium", but this was not the powerful substance they were seeking. Slowly though, they were beginning to isolate that substance and by the end of 1898 they were able to christen it "radium". Yet the fact remained that no one had seen it, nobody knew its atomic weight. To prove its existence, Pierre and Marie were to work for another four years. The pitch-blende, found in Austria and used in the manufacture of glass, was expensive, impossibly expensive, but they reasoned that the residue after glass manufacture would be suitable for their purpose. They persuaded the Austrian Government to allow them a ton of the stuff, if they would pay for its transport to Paris. There was a long delay. Then one morning a horse-drawn wagon drew up and delivered a load of sacks, full of the dull brown ore, still mixed with the pine needles of Bohemia.

Painfully over the weeks they discovered that their guess of one per cent had been ludicrously wrong. The radiation was so powerful that only the tiniest quantity of radium was needed to produce amazing phenomena—but they seemed no nearer to isolating it. "Pierre," Marie said one day, "what form do you imagine it will take?"

"I don't know. But I'd like it to have a very beautiful colour." By now Pierre would cheerfully have abandoned the task of preparing pure radium. What did it matter? Surely the meaning of the phenomenon was more interesting than its material reality? He urged Marie to give up, but he was unable to shake her determination.

Forty-five months after the day on which they had announced the probable existence of radium, Marie was successful. She prepared a tenth of a gram of pure radium, measured its atomic weight. At last the element existed officially. It was more beautiful even than Pierre's "beautiful colour": it shone by itself, like a glow-worm.

Radium had been isolated, but there was much to do; because they had devoted four years to the task when Pierre might have been advancing himself in more profitable research, they were desperately poor. They refused an offer to work as a team at the University of Geneva, as this would have interrupted their researches on radium, which was almost impossible to transport. In order to keep body and soul together, Pierre accepted a junior position teaching children.

Life was bleak, almost hopeless it seemed, until the exciting discovery that radium destroyed human cells. It could thus, under proper conditions, be used to destroy diseased ones; it might cure growths, even cancer. At last—and the Press was quick to take up the tale—the new element could be shown to be useful. A French industrialist founded a factory to make it, gave the Curies a handsome laboratory. Then a personal problem arose: now that radium would soon be manufactured on a larger scale, perhaps all over the world, they ought—it was simple common sense, their friends said—they ought to patent the process. In this was they would become rich.

The Curies refused to take out a patent. Radium, they declared, belonged to the world; no one had any right to demand a profit from it.

Though they were not rich, honours poured in on them. In 1903 they shared the Nobel Prize for physics with Henri Becquerel and from then on their life seemed a series of prize-givings. A sudden demand sprang up for radium; journalists descended on them, discovered or invented details of their private life, hazarded wild guesses as to the availability and curative properties of the new element and succeeded in arousing impatience and anger among the public: why was the new panacea not available, to all, what were the Curies doing? Once again, life became difficult, brightened only by the birth of a second daughter, Eve, when Marie was thirty-seven.

Tragedy struck in 1906. Pierre was run down by a cart in the street and killed. He was by this time director of the physics laboratory at the Sorbonne and in that post Marie succeeded him, the first woman ever to be appointed. Her researches continued. In 1911 she received her second Nobel Prize for chemistry.

Throughout the First World War Marie Curie, refusing to flee from Paris, organized and equipped X-ray stations and X-ray care for the wounded. Famed as the discoverer of radium, she also knew more than anyone else about the application of the X-ray, and she put her knowledge, selflessly, to use. Thousands of wounded passed through her hands. Governments sent representatives to her "Radium Institute" to learn the uses of X-ray, of radium therapy. At the end of the war she had the double satisfaction of peace and a free Poland. Her native land, slave for a century and a half, became free again.

Marie Curie was the first known victim of radioactivity. The disease from which she had suffered for so long had been variously diagnosed as bronchitis, cancer and tuberculosis: eventually it was proved to be "pernicious anaemia, caused by the destruction of bone marrow after a long accumulation of radiations". Marie Curie died in July, 1934, in the service of mankind.

Robert Goddard

(1882–1945)

Father of the Space Age

Robert Hutchings Goddard, who invented the space rocket, was born in Worcester, Massachusetts, USA, on October 5, 1882. When he was less than a year old, the family moved to Boston. His father, Nahum Goddard had opened a small machine tool factory in Boston, which prompted the change.

Bob (short for Robert) learned to read at an early age and liked books and magazines that had to do with science and inventions. He wanted to know how everything worked. He would spend hours examining the wheels and springs inside a clock or inspecting the mechanism of his mother's sewing machine.

The sky fascinated young Robert. He built kites to explore it and flew them in all kinds of weather and for hours at a time. He wanted to fly up there. But at that time there were no aeroplanes for him to do so.

Bob lived in Boston for the first sixteen years of his life. He was an only child. He and his parents were happy together in their comfortable big house. His father's factory was doing well. Bob liked his school and classmates. His father was a jolly, cheerful person whose whistling could be heard a block away when he came home in the evening. Their home was full of gaiety and laughter.

Soon however, sadness and worry arrived, as Bob's mother became ill with tuberculosis. It was a dreaded disease in those days for no medicine had yet been discovered to cure it.

His mother's illness took a serious turn and the doctor advised complete rest and good clean country air. So the family moved back to Worcester, forty miles away, to live with his grandmother in a spacious old farmhouse at the edge of the town. It was the house where Bob had been born and where he spent most of his summer vacations.

The move to grandma's house helped his mother a lot. Within weeks she began to look and feet better. But Bob fell sick with stomach trouble and a severe kidney ailment. The family doctor advised complete rest for a year.

So Bob stayed home and rested till he was tired of resting. He became better by late autumn and was permitted to go out. He took to walking in the orchard for short periods. After a while Bob was able to rake up fallen leaves and do other outdoor work. That was when a dream came to him — an amazing dream that was to guide and shape his whole life.

It was a daydream and it came to him in the afternoon of Oct. 19, 1899. He had turned seventeen about two weeks back. He had gone out and was walking in the orchard. Noticing that a tall cherry tree needed pruning he climbed it. He was cutting off the small dead branches, resting every now and then, for he was still weak from his illness, when it came.

Bob began to imagine that he was in some sort of flying machine. The machine was being sent up into the sky by a secret device that allowed it to escape the pull of the earth's gravity. The daydream was so real that he could almost see the machine soaring into outer space. He didn't know how long the dream lasted but when he climbed down the tree he was, as he wrote, a different boy. He was now certain that his life's work would be the exploration of space.

Though Bob's health improved considerably during the year he was out of school, he was not well enough to satisfy the family doctor. So he was advised to remain at home for another year.

He rejoined the school in the autumn of 1901. When he reached the senior class he was twenty-one and most of his classmates were still boys. So, it is not surprising that they elected him their class president. He was also elected by teachers to give the oration at the graduation function. He ended his oration with the thought: "The dream of yesterday is the hope of today and the reality of tomorrow."

To pursue this dream he joined the Worcester Polytechnic Institute the following September.

Bob graduated with a B.Sc. degree from the Polytechnic on June 11, 1908. He wanted to enter the university for higher studies, but the illness of his mother had cost much money and he wanted to spare his father more expenses. So he secured the job of an instructor at the Polytechnic.

A year went by. Bob had saved enough money to join the Clark University. He took his master's degree in 1910 and one year later his Ph.D. He was now Doctor Robert Goddard.

He arranged to stay at the university so that he could use the laboratory for his experiments. He began working on rockets. He was continuing this when a great honour came to him, in 1912. He was offered a research fellowship at Princeton University and he jumped at the chance. There he worked harder than ever in his eagerness to find out the secrets of space. He spent all day in the lab and worked late into the night too.

When he went home to Worcester for the Easter vacation he was feeling exhausted and miserable and running a temperature. The medical examination showed that he had tuberculosis of lungs. The doctors didn't tell him then but he found out later ... he had been given only two weeks to live.

But Dr. Goddard refused to die. "They aren't going to write me off yet," he said when he heard the medical opinion. "I'm going to get well." And indeed he made slow but steady progress in his fight for health. By the autumn of 1914, he had recovered enough to teach physics part time at Clark University and resume his rocket experiments.

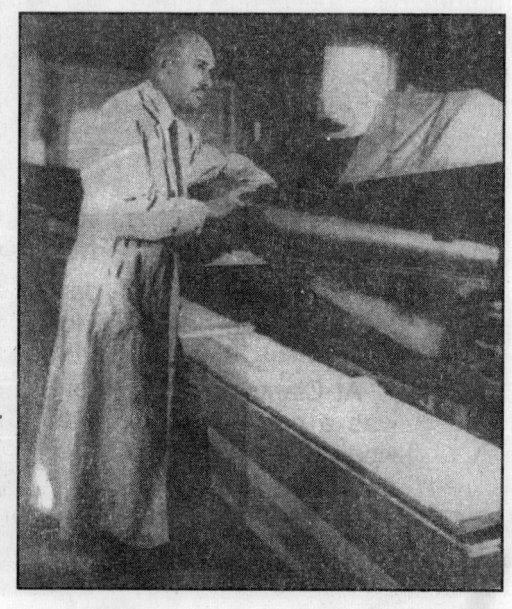

There was one experiment in particular that Bob wanted to try. He wanted to prove that a rocket could operate in the vacuum of outer space. In those days even scientists believed that such a thing as space flight was impossible. The theory that a rocket could fly through space where there was no air for it to push against was not taken seriously. Dr. Bob based his thinking on a principle expounded centuries ago by Sir Isaac Newton, in his third law of motion: *For every action there is an equal and opposite reaction.* He was sure that the action of the gases escaping from a rocket's burning fuel caused a reaction in the other direction, by propelling forth the rocket in space. "I'm going to reach the stars, Ma," he told his mother one day. "And I'll bring one back for you."

At Clark University there was a pretty girl who was working as a secretary. Her name was Esther Kisk. She was working and saving money to go to college. Dr. Bob wanted someone to do typing for him and took in the smart and bright Esther. They became close friends and fell in love. The fact that Esther was many years younger to him did not matter to either of them and finally on June 21, 1924, they were married.

For five years, from 1920 until 1925, Dr. Bob designed and built a number of rockets with liquid fuel motors. Then, he launched the world's first liquid fuel rocket on March 16, 1926. The flight had lasted less than three second. The distance covered had been no more than 184 feet. The speed had not been much more than sixty miles an hour. Yet on that cold spring day history had been made. The day was to become as important as December 17, 1903, when the Wright Brothers made the first heavier-than-air-flight.

Further tests were made with improved designs, and on May 31, 1935 the rocket reached a height of 7500 feet. But what was really amazing was the fact that this rocket travelled at the incredible speed of 700 miles an hour which is very close to the speed of sound. Mankind had reached the threshold of the Space Age !

When the Second World War started and the United States entered the war, Dr. Bob closed his laboratory and left his testing grounds. He started working on a type of rocket motor for the U.S. Navy. But his health was failing.

In the summer of 1945 he had to be hospitalized for a serious throat operation. The operation was successful and after a while Dr. Goddard was able to sit up and scribble some notes about his future work.

But suddenly he took a turn for the worse and on August 10, 1945 at the age of 63, Dr. Bob Goddard, the genius, passed away.

Many posthumous honours were bestowed on Dr. Goddard by America. Esther Goddard, his widow, was presented with the country's highest decoration for civilians, the *Congressional Gold Medal* and with aviation's greatest award, the *Langley Gold Medal.*

Great research laboratories like the Goddard Space Centre at Greenbelt. Maryland, and the Goddard Institute of Space Study in New York City were named after him.

The US House of Representatives, on July 2, 1962 established a day of national honour in memory of Dr. Robert H. Goddard. The date is March 16, the very day when in 1926, Dr. Goddard launched the world's first liquid-fuel rocket.

Reuven Feuerstein, Dr.

This Israeli psychologist's controversial "brain-gym" exercises are teaching thousands of mentally handicapped youngsters around the world to speak, write and reason.

At the age of 16, Alex. struggled to hold a fork to his mouth. He had a very low IQ, and a rudimentary vocabulary. Letters and figures were meaningless to him.

Doctors said it was unlikely that he would ever significantly improve, for he had only half a brain. The left hemisphere had been surgically removed when he was eight in an attempt to phase out the drugs that he was taking to control his epileptic fits.

Specialists in England, Alex's home country, predicted he would never really learn to read or write. Then Alex's mother heard of Professor Reuven Feuerstein (pronounced Foyerstein), an Israeli psychologist who had worked miracles with apparently hopeless cases. Alex spent two years with him in Jerusalem.

I met Alex last year. At 19, he sported a mane of blond-dyed hair and greeted me with a cheerful handshake.

I handed Alex a news I'd saved from a magazine. "Would you mind reading this for me?" I asked him.

"Right!" He put on his reading glasses. *"When you meet him,"* he began, *"Alex does not strike you as a particularly special or unusual boy. He has a birth mark on his head that looks just like Gorbachevs....*" He smiled, recognizing himself.

In Jerusalem, the boy with half a brain had learned to read. He had also learned to write and do basic maths.

Feuerstein's pioneering work for deprived or mentally handicapped children has helped thousands of kids written off as hopeless to live fuller, sometimes even normal lives.

In a classroom in downtown Jerusalem, I watched Alex and two other teenage Feuerstein students work with "Orientation in Space," one of a set of "brain-gym" exercises devised by the professor.

On the table before them was a picture of a square. Seen from a faceon perspective, it had a house on the far side, a tree to the right, a flower bed to the front and a bench to the left.

Ruthie Kaufman, one of Feuerstein's instructors, now asked Alex to mentally transport a boy into the middle of the square. The imaginary boy was facing right.

"Problem," said Ruthie. "What does the boy see in front of him?"

"A tree," Alex answered at once, correctly.

Next the boys took it in turns setting the problems. They got progressively harder. "If the tree is to the boy's left, which way is he facing?" one of them asked.

"He's facing the house," Alex answered, wrongly (the boy would have been facing the flowers).

"How come you got it wrong?" Ruthie asked him.

"I didn't think," Alex replied. "I wasn't focused."

Like other Feuerstein exercises, "Orientation in Space" helped students to acquire a number of basic thinking skills, Ruthie told me afterwards. It taught them about points of view and how to adopt problem-solving strategies. Think before you speak was one. Check your work was another.

Reuven Feuerstein was the son of a Romanian rabbi. When the boy was eight, an illiterate coachman asked Reuven to teach him to read the Bible. In payment, he promised Reuven his pocket watch.

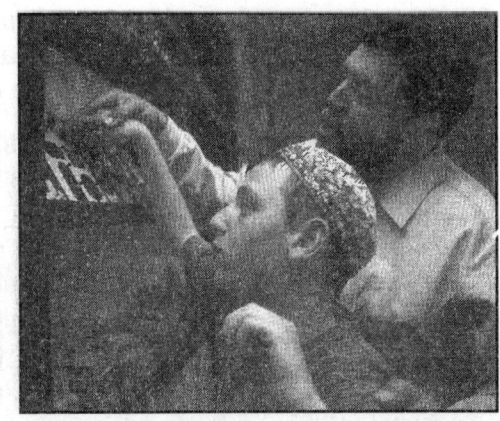

"He learned to read, but I never got the watch," laughs Feuerstein, who is now 80. "It's the story of my life." For Feuerstein has always made it a point not to charge his patients, and his International Centre for the Enhancement of Learning Potential (ICELP) in Jerusalem survives largely on voluntary donations.

At the start of the Second World War, Feuerstein was a 19-year-old psychology student in Bucharest. He became involved with the Jewish underground, helping to smuggle to Palestine children whose parents had been sent to concentration camps by the Nazis.

In 1944, he was put on a boat to Tel Aviv. After the war, he set up two camps for young Holocaust survivors who had suffered traumas and performed poorly at school. In 1948, he was sent to recuperate from tuberculosis in Switzerland, where he later studied under Jean Piaget, considered by many to be the father of developmental psychology.

Back in Israel, he went on to develop the "Feuerstein Method" for helping low performers or retarded youngsters. It offers 14 sets of exercises which aim to improve intelligence and the ability to learn. It has been translated into some 17 languages and is currently being taught in about 55 countries.

I watched Daniel, an 11-year-old from New York, consider a jumble of dots on the page before him. Daniel, diagnosed as having a low IQ and being incapable of prolonged attention, used his pencil to join up four of the dots into a square, and three into a triangle. Absorbed, he moved on to the next puzzle.

"Organization of Dots," usually the first exercise in the Feuerstein curriculum, helps children with an often sketchy grasp of reality to order the chaos in their minds. The exercises become progressively more difficult, as I found out when it took me ten minutes to solve an intermediate problem.

Feuerstein sat besides Daniel, encouraging him with a heartfelt "Bravo" every time he got a problem right.

Many educators believe handicapped children should only be assigned undemanding, generally manual, tasks. Feuerstein believes in challenging his pupils with complex intellectual problems. After the dots come exercises that hone analytical and organizing skills, and promote the capacity for abstract thinking.

"Even when damaged, the brain is able to regenerate," he told me. "Low intelligence is a state, not a trait. States can be altered. I want to help mentally-impaired children modify themselves so they can live a full life."

ROBERTO was a good-looking boy who had been diagnosed as autistic back home in Costa Rica.

Tracy Stevens, a young ICELP instructor, sat with Roberto at a table. She drew a square. "This is a square," she told him, speaking slowly. "It has four sides."

She drew another, bigger square next to it. "Are the two squares the same?" she asked.

Roberto's gaze was on the ceiling. Tracy reached out and laid her hand on his, palm-to-palm. "What's the difference between our hands?" she asked.

"Mine's smaller," Roberto mumbled.

"Good!" exclaimed Tracy. "Now, what's the difference between the squares?"

As the lesson progressed, Roberto's gaze stopped wandering. He answered more willingly, and was right more often. If his answer was wrong, Tracy simply repeated or rephrased the question. If he answered correctly, she praised him.

Tracy drew a hexagon. "How many sides does it have?" she asked him.

Using the tip of his pencil. Roberto counted off the sides. "Six" he answered correctly.

It was difficult to believe this was the same boy of half-an-hour before. Using squares and triangles, Tracy had introduced him to the idea of differences and similarities, to the notion of relationships—a basic tool we use to order and understand the world around us.

I accompanied Feuerstein on a three-hour drive to a kibbutz in the Negev region, where two of his ex-pupils, both with Down's syndrome, were getting married.

The groom worked on the kibbutz farm and wrote poetry; the bride was a kindergarten assistant. They threw their arms round Feuerstein in obvious delight.

His track record with children with Down's syndrome is striking: Feuerstien has shown that they are fully capable of becoming self-supporting.

"Before I went to Israel, I couldn't talk about my feelings. I had no name for them," Peetjie Engels, a 23-year-old Dutch girl, told me when I visited her in the Netherlands. A child with Down's syndrome, she could communicate only through disjointed words. She started training with the Feuerstein when she was nine.

"Working with the dots taught me to concentrate and to deal with problems systematically," she said. Even more important, it helped to open her mind to the world of abstract thought. Since trial and error with a pencil was banned, she was forced to draw "virtual lines" between the various points on the page, which, joined together, made "mind shapes."

Peetjie went on to discover that an apple is similar in shape to an orange, but different in taste. Now she saw that some things could be the same, yet different. She is training to become a kindergarten assistant and lives alone close to her parents' home.

Increasingly, the Feuerstein method is finding applications beyond its original target group.

Avi, an Israeli army soldier, suffered extensive brain damage after being shot in the head by a sniper some five years ago. When he was sent to Rashmi Sharma, a young Indian psychologist at ICELP, his vocabulary was down to a few words.

"We had to recreate his ability to use symbols so he could learn to read again," said Dr Sharma. Avi can now read, and his vocabulary has grown significantly.

Impressed with this and other success stories, the Israeli government has signed a contract with ICELP to rehabilitate military personnel who suffer head injuries.

A number of corporations, such as Motorola in the United States, have also experimented with the Feuerstein method to enhance executives' performance.

One of Israel's largest corporations, Israel Aircraft Industries (IAI), went to Feuerstein to improve its employees' ability to learn. After 60 hours of training, IAI technicians could diagnose and remedy aircraft problems faster and more accurately, and a higher percentage of them achieved the civil aviation maintenance licence.

In schools around the world, the Feuerstein method is helping non-handicapped students; Brazil's populous state of Bahia, for example, is introducing the Feuerstein course into its high school curriculum. The biggest such programme ever launched, it will, by completion in less than three years time, be taught to 600,000 students in 500 schools.

The Feuerstein method is still viewed with suspicion in some academic circles. But one of the world's foremost authorities in the field, Robert Sternberg, IBM Professor of Psychology and

Education at Yale University, comments: "Feuerstein will be remembered as one of the truly major figures of the 20th century in the human-abilities movement. He has made a remarkable contribution to our understanding of abilities and their modifiability."

AT 8:15AM on a December day in 1988, Reuven Feuerstein entered a lecture room at Bar-Illan University. "Congratulate me. I am proud to announce I've just become a grandfather," he told his students. "My grandson has Down's syndrome."

He says now: "I'd told countless parents that the birth of a baby with Down's syndrome was cause for joy, not sadness— and then, when my grandson Elchanan was born, it happened to me. I'd often wondered how I would react. But I passed the test. Elchanan is a source of joy to me."

I watched the man who has changed the lives of thousands of children head off down a corridor. He held a small figure by the hand—Elchanan, now 12. Feuerstein has taught him almost since birth, and the boy is doing well in a regular school.

"He will grow up to marry. He'll have a regular job," said his grandfather.

"What kind of job?" I asked.

"A teacher, may be."

Robert Watson-watt

(1892–1973)

The Radar Man

In what is now known as the "Battle of Britain" the German Luftwaffe and bombers raided England, day and night, in overwhelming numbers in 1940. Although England had comparatively fewer Spitfires and Hurricanes, she still managed to throw back the massive air attacks. With due respect to all those overworked airmen who destroyed the German air strength, it cannot be denied that they could not have done so without the help of a secret weapon. Before the German aircraft could even cross the sea to attack England, this weapon was able to inform the Royal Air Force and Civil Defence of the strength and targets to be attacked by the enemy aircraft. Even Adolf Hitler paid reluctant tribute to the weapon when in 1943 he remarked in a speech that a secret weapon invented in England had foiled Nazi Germany from achieving its war aims. That weapon was the Radar and its inventor was a Britisher, Robert Watson-watt.

Son of a Scottish carpenter, Watson-watt was born in Brechin, Scotland, on April 13, 1892. From the childhood he had an inventive bent of mind and was keen to become an engineer. He took a degree in electrical engineering but after teaching science for some time he felt he was more interested in the applications of science. So, when he was offered a job in the Meteorological Office of the Royal Aircraft Establishment at Farnborough, he gladly accepted it. This was a surprising turn of events for him - and for good - though at that time meteorology was not even considered to be a science. But as Watson-watt enjoyed the work he stuck on to it. In those days, wireless telegraphy - the technique that uses radio waves to send and receive messages - had a recurring problem. All kinds of noises - crackling, hissing, etc, - were heard over the wireless telegraph sets. These noises were called "Atmospherics" because their source was believed to be in the atmosphere, especially the electrical phenomena, such as lightning flashes in thunder clouds.

As atmospherics were a nuisance to telegraph set operators, the authorities of the Meteorological Office were keen to know if the onset of thunder clouds could be predicted in advance, so that operators would be prepared to face them. Watson-watt thought that there was no point in this investigation unless he was sure about the connection between atmospherics and thunder clouds. He therefore decided to study systematically the connection between these two phenomena. He shifted to the nearby bigger meteorological station at Aldershot and began to perform his experiments. In those days, it was a well known fact that when the aerial of the receiving telegraph set was perpendicular to the incoming radio waves, it gave the loudest sound signal. Thus if two aerials were employed to catch a radio signal at its loudest, the point at which the perpendiculars to the aerials crossed each other gave the location of the radio signal. The British Post Office used to employ this technique to track and catch unlicensed radio sets. Even during the World War I, the British Admiralty was able to track down the invading German Fleet sailing towards the site of the battle of Jutland in a similar manner. Watson-watt wondered whether he could use this technique of radio location for tracking atmospherics and thus thunder clouds. He drew up a big programme - the biggest till then conducted - in which he sought the assistance of the British

Broadcasting Corporation and a large number of meteorological observers and telegraph operators. The area of study was even beyond the border and coastline of England.

In Watson-watt's programme, the British Broadcasting Corporation broadcast certain talks and the telegraph operators were asked to report when atmospherics occurred during the course of the talks. Simultaneously, observers located in and around England watched for thunder clouds and lightnings. Using two aerials, Watson-watt and his team located the source of atmospherics. Through this he deduced that atmospherics were indeed produced by lightings in thunder clouds. Using what he called "radio location" technique, he could track lightning that occurred as far as 7,500 kilometres away. It was a major achievement for 35-year-old Watson-watt and the authorities recognised his meritorious work. In 1927 he was made the incharge of the Radio Research Station at Slough, Buckinghamshire. Here he conducted extensive research on what is today known as "Ionosphere", a term coined by him for the electrical layer surrounding earth, which reflects radio waves. Subsequently, he was appointed the head of the Radio Division of the National Physical Laboratory at Teddington.

It was here in 1935 that an unusual official inquiry came to him. The Air Ministry wanted to know whether it was possible to develop the proverbial "Death rays" in the laboratory. In those days, death rays were believed to have the power to kill men, stop running vehicles and trigger off ammunitions from a distance. It was rumoured that Nazi Germany had already developed this deadly weapon, which was likely to give her an edge in the war expected in near future. Watson-watt replied that "Death rays" were "not a good bet". Instead, he suggested that the invention of a device which could detect enemy aircraft was then possible. The device could function in both day or night and in fog and clouds. The device could detect an aircraft using radio waves. To his own surprise, the authorities asked him to give an immediate demonstration of the device, even though it may by a crude one. The fear of the impending war had prompted the Air Ministry to take this decision. And Watson-watt made immediate preparations.

Watson-watt thought that the wings of an incoming aircraft would reflect transmitted radio waves like a mirror, which could be immediately caught by telegraph set to indicate the presence of the aircraft. It was a very simple concept and proved to be equally successful in practice. Using the powerful radio transmitter of the British Broadcasting Corporation at Daventry, radio waves were beamed at an incoming aircraft and the reflected waves were promptly detected. This preliminary demonstration which took place on February 26, 1935, showed the Air Ministry what it was expected to show—the promise and potential of the device in the forthcoming war. Immediately, the device became a war secret. Though all through the Second World War, it was extensively used, nobody knew who had invented it. Watson-watt came into limelight only after the war when all behind-the-scene details of the war were revealed to the public. He then became a national hero overnight.

In the beginning, only Watson-watt knew the importance and potential of his device which he called "Radio-location". Every day, his equipment mounted on a truck was taken to a lonely spot and brought back to analyse the data collected. When some people saw these strange goings-on on the coast of Suffolk, they thought that "Death rays" were being developed. In due course, Watson-watt's team knew what he was doing as the equipment for detecting aircraft was being perfected. In fact, a time came when thousands of persons were trained in the art of detecting aircraft. In December 1935, five stations exclusively for this purpose were installed in some remote spots on the east coast of England. In the meanwhile, the range of detection of Watson-watt's device also increased. By the middle of 1936, it could detect an aircraft 120 kilometres away. In 1937, 20 more such stations were installed along the coast and by 1939 a chain of such stations from the coast of Scotland to Isle of Wight had been installed. All were ready to detect the presence of an invading German aircraft expected to arrive from that direction. Britain was well prepared for the war, thanks to Watson-watt.

In the autumn of 1939, the first German aircraft raids began. The aircrafts were detected long before they reached the coast of England. Even the height and range of the aircrafts were ascertained using Watson-watt's device. Anti-aircraft gunners could therefore shoot at the aircraft still invisible to their eyes. In fact, during the Battle of Britain, the gunners were able to shoot hundreds of enemy aircraft which the device showed them. In due course, Watson-watt also invented a device which could tell a gunner whether the incoming aircraft was a friend or an enemy. A device was installed aboard all the British aircraft. It emitted a particular radio message which indicated the friendly presence of an aircraft. When the United States entered the war, Watson-watt used to frequently visit that country to advise the technical experts there on matters regarding his device, which had in the meantime come to be called *"RADAR"* for *"RAdio Detection And Ranging."* In 1942, he was even knighted by the British Government for his valuable services during the war. However, for security reasons, nothing clear about his contribution to the war was mentioned. In the meanwhile, many other applications of radar, both offensive and defensive, were discovered and promptly brought into use. Radar was installed aboard aircraft to detect warships on the seas and to help night bombers in locating their targets.

Although radar was developed as a secret war weapon, by the time the war ended many of its potential peaceful applications had been discovered. In fact, Watson-watt was very happy when his device began to find use during peaceful times. Douglas harbour in the Isle of Man was the first one to be equipped with radar to detect ships and aircraft in dense fog. Thereafter, the peaceful applications proliferated. Nowadays, a radar is used in airports and shipyards. It is also utilised in understanding and predicting weather - the aim with which Watson-watt began his experiments. Radar astronomy has also come up as a science to probe deeper into the mysteries of the heavens. Recently, radar has also been installed aboard spacecraft to map the surface of nearby planets. Could Watson Watson-watt have imagined all these applications when he put up his proposal to locate aircraft by radio waves? No, never. Such are the unexpected gains of an invention.

In 1949, Watson-watt retired from his Government job and started his own company. Much later in 1973, when his invention had already become a part and parcel of every day life, he died. He was one of those few privileged inventors to have seen the wide acceptance of his invention in his own lifetime.

Phased Array Radar

For any one who has visited an airport, the revolving and steerable concave-shaped antenna of a radar is a familiar sight. But the day is not too far away when we would be missing this familiar sight. One may wonder how, for a radar is an indispensable part of an airport. A radar would certainly continue to be a part of a future airport, but it would not be as conspicuous an installation as it is today. It would simply be a disk, one amid various other gadgets atop an airport building, but it would be more powerful and efficient and would also track several moving aircrafts simultaneously in almost all directions. This new type of radar would be *"Phased Array Radar"*. Due to its inbuilt advantages over the present radar, it would be installed in space not only to keep an eye on enemy targets but also to guide sea traffic, watch movements of icebergs and to map the resources of earth at a faster rate. The phased-array radar is able to sweep the sky from horizon to horizon without any physical rotation and track moving targets or objects almost simultaneously. The biggest advantage of this system is that the radio beam could be steered from one direction to another within a matter of a few microseconds. The added advantage is that the system would be in operation 'though partially' if in case one of its radio sources goes out of order due to a blown up tube, unlike conventional radar where the blowing up of a single tube would render it inoperative.

At present, two U.S. military radars which are phased array systems are COBRA DANE and PAVE PAWS. The COBRA DANE installed on the edge of the Bering Sea to observe the Russian Ballistic Missile tests consists of 15,360 radio sources on its disk. It can detect a metallic object, the size of a grapefruit, at a distance of 1,800 kilometres! The PAVE PAWS is a chain of phased array radars installed on the Cape Cod in California and also on sites in Georgia and Texas. It is for watching the sea-launched ballistic missile. The PAVE PAWS is however more reliable than the COBRA DANE because it is totally transistorised. In near future, the phased array radars would also be used at airports for keeping an eye on the aircrafts approaching the runaway from all directions.

Salim Ali

(1896–1987)

Pioneer Bird-Man

In the 19th century, the idea of nature conservation was virtually unknown. To the British rulers, hunting was a form of sport. Well-known hunters used to visit this country attracted by the profusion of wildlife. When these veteran hunters went back, they wrote books describing their exploits. Two or three rhinos or tigers before breakfast was hardly an exaggerated claim for a trigger-happy enthusiast. Remarks like these found in those memoirs indicate how rich in forests and wildlife India must have been in those days.

It was an age when hunting for sport was not only considered manly but also a status symbol for the upper middle-class Indian. Salim Ali was born in such a family. Having lost his parents very early, he and his eight brothers and sisters were brought up by a kind and loving uncle. The home atmosphere shaped young Salim and prepared him for his future career. His uncle, Amiruddin, was an expert hunter and on very friendly terms with the sport-loving princely community. He was frequently invited to their hunting parties. Salim used to admire his uncle, who presented the boy with an air gun when he was nine. Salim grew up wanting to be an explorer and a big-game hunter. In his childhood, he was used to outdoor life. With his cousins, he roamed about the countryside shooting birds. In one of these outings, when he saw a dead sparrow with a yellow streak on its throat, his interest in the study of birds was aroused.

A common sparrow does not have a yellow streak on its throat. Salim was curious. He consulted his uncle who happened to be a member of the Bombay Natural History Society (BNHS). With a letter of introduction from him, young Salim went to meet the secretary of the society, W.S. Millard. Mr. Millard identified the bird as a yellow-throated sparrow. Moreover, he took the young boy round the society's collection of specimens. A new door had opened. It was a decisive moment in Salim Ali's life. To quote his exact words: He patiently opened drawer after drawer for me to see the hundreds of the different birds found in the Indian empire, and I believe it was at this moment that my curiosity about birds really clicked.' (*The Fall of a Sparrow*)

Good books on Indian birds were not freely available those days. Mr. Millard offered the use of the society's library to the young enthusiast. He also encouraged the boy to develop his hobby in a more scientific manner by proper collection and preservation of specimens. All this was very exciting for young Salim. He eagerly went through the books in the library but he found that they were not very comprehensive. In fact, there was no systematic study of Indian birds. It was left to Salim Ali to make up for the gaps in knowledge about bird life in India.

Salim was not good at his studies. The family had a business in Burma, and he was sent there at the age of eighteen to assist a cousin. Soon Salim began to realize that he should complete his formal education. Returning to India, he took a one-year course in commercial law and accountancy. At the same time, he completed the B.Sc. course in zoology from St. Xaviers College, Bombay.

Since he was going back to Burma and perhaps would not get a chance to return to India soon, his sisters insisted on his marriage. So, at the youthful age of twenty-two, he was married to Tahmina. She had been brought up in England but had the remarkable ability to adapt herself to the different situation she now faced in India. It was not just that her husband did not have a regular job, but also that his passion for birds took him to out-of-the-way places. Tahmina was soon infused with her husband's love for ornithology.

When Salim Ali returned from Burma, he tried his hand at various jobs. For some time, he worked in the Natural History section of the Prince of Wales Museum. He was already convinced that ornithology as a full-time occupation was the only thing he was interested in, but he did not have the professional training of a field ornithologist. There was no place in India where he could get such training. So he wrote to the Zoological Museum of the Berlin University. He had also written to the British Museum in London, but they did not seem very keen to take him. However, the response from Berlin was encouraging, so he decided to go there. Berlin proved to be, in his own words, *'the luckiest turning point'* in his ornithological career.

He returned from Germany in 1930. There was still no regular job for him. He did not know how to utilize his expertise. It was then that the idea of undertaking regional bird surveys came to him. He offered his services to the Bombay Natural History Society, provided they paid for the expenses of the surveys. Salim Ali did not want any salary. This was how the different ornithological surveys began—the first one being in the princely state of Hyderabad.

Bombay was an expensive place. It was becoming increasingly difficult for Salim Ali to maintain his family, since he did not have a regular job. So they decided to move to Dehra Dun from where he would have an opportunity to study the birds of the mountains.

The years spent in Dehra Dun were very fruitful ones but this stay was tragically cut short by Tahmina's death in 1939. Once again, he returned to Bombay. This move was partly prompted by a request from close relatives and was partly because of the BNHS, which provided excellent facilities for his work. All his life, his close relatives stood by him, providing emotional as well as material support. This is what sustained him through the ups and downs of life. In his autobiography, *The Fall of a Sparrow*, he has confessed how much he owed his family where his development as a naturalist was concerned. In the same book, he recalls with gratitude the help and encouragement he got from his elder brother, Hamid. However, not all his elders approved of his obsession with birds. His maternal uncle, Abbas Tyabji, a judge of the Baroda High Court, always thought Salim should be engaged in more useful work. He considered bird-watching an excuse for indolence. He died around 1935 and Ali wrote of him, 'It is pity he did not live to see that the indolence has paid dividends.'

Salim Ali made a series of bird surveys in Travancore, Bahawalpur, the Eastern Himalayas, even in Afghanistan and Tibet. In 1934, he drew up a project for research in economic ornithology. This included scientific studies in the feeding habits of birds and their bearing on agriculture and forestry. He was full of creative ideas, but the question of finance proved to be a hurdle at the early stage. But slowly, as his work gained recognition, doors opened and he plunged into a life devoted to the study of Indian birds, thus becoming a pioneer in this field.

Salim Ali has written a number of books on Indian birds—of Kutch, Kerala, Sikkim and other places. His famous *Book of Indian Birds*, published in 1941, has run into many editions. His greatest work is the ten-volume *Handbook of Birds of India and Pakistan*, jointly written with Sidney Dillon Ripley of the Smithsonian Institute of Washington. They had first met in Bombay in 1944. They became lifelong friends. Their joint venture spanned over decades. Salim Ali had to spend some time in the United States doing library research. The first volume was published in 1968, and the last in 1974 when Salim Ali was seventy-eight years old.

He died on 20 June, 1987, at the age of ninety-one. He was active to the last, playing a significant and decisive role in wildlife conservation in India. It was through his timely intervention that the unique bird sanctuary at Keoladeo Ghana in Bharatpur was saved. This used to be a private duck-shooting preserve of the Bharatpur rulers. Even after Independence, the rulers of this state refused to give up their shooting rights. Salim Ali was able to convince Nehru of the necessity of protecting this wetland and the sanctuary was eventually saved.

Salim Ali persuaded the Bombay University to recognize the BNHS as a centre for postgraduate research in field ornithology, leading to master's degrees and doctorates in zoology. He received many awards but above all, he evoked the spontaneous love and admiration of his countrymen. J.B.S.Haldane always gave the example of Salim Ali for doing the impossible with hard work, dedication and just a pair of binoculars.

Samuel Hahnemann, Dr.

(1775–1843)

Founder of Homoeopathy

Homoeopathic system of medicine was developed by Hahnemann. In this system infinitesimal amount of medicine is given to the patient. Homoeopathy has grown into a system with a place of its own. Most of us, at one time or the other, have benefitted from the curative powers of homoeopathic medicines after not getting relief from other systems of medicines. Homoeopathic medicines are dispensed in tiny sugar pills doused with a few drops of alcohol containing tiny amounts of medicine. Efficacy of Homoeopathic medicines has been established beyond any doubt.

Atomic theory of matter, on the other hand, was developed by John Dalton (1766-1844). Its basic premise is that matter consists of small particles called atoms. Atoms of different substances such as iron, gold are different. An atom, it was discovered later, consists of a heavy nucleus containing positively charged protons and uncharged neutrons. Negatively charged electrons, equal in number to the number of protons, revolve around the nucleus in different orbits. This is known as the structure of an atom and different atoms have different numbers of protons, neutrons and electrons.

When homoeopathic system of medicine is analysed on the basis of the atomic theory of matter, contradictions appear. Homoeopathic medicines are prepared by successive dilutions of the original salt or solution in alcohol. These medicines are given in centesimal potency, denoted by putting numbers such as 1, 3, 6 etc after the name of the medicine, or in decimal potency, denoted by adding the numbers such as 1X, 2X etc after the name of the medicine. For preparing first centesimal (reckoned by hundreds) potency medicine one part of the original medicine is mixed with 99 parts of alcohol thoroughly. If one part of one centesimal potency medicine is mixed further uniformly with 99 parts of alcohol one gets 2 centesimal potency medicine. This process repeated 30 times with successive dilutions gives 30 centesimal potency medicine and so on. Decimal potency medicines are prepared in a similar manner but dilution is one to ten. Clearly, 1 centesimal potency medicine is same as 10X potency medicine.

A simple calculation shows that beyond 12 centesimal potency likelihood of a single atom of the original medicine in the successively diluted form of the medicine is extremely small. Keeping in view that centesimal potencies 30, 200 or higher are commonly prescribed and proved effective, a dichotomy arises. If the higher potency medicines do not contain even a single atom of the medicine, where does the curative power of the medicine come from? Not much attention has been paid to this basic question of homoeopathy defying the atomic theory of matter. If one attempts going deeper into this question, obviously one has to go beyond the atomic theory of matter. Since understanding beyond the atomic theory is still at an elementary level at best one can resort to guarded speculations. It can be argued that atomic theory is a gross version of some finer (but still not discovered) structure of matter. Another way out of this paradox is to assume that an atom during its stay in a liquid modifies it in some subtle manner. This modification is atom specific and difficult to detect with available instrumentation. Is this modification the footprints the atom has left behind in the liquid? With the present understanding of the atomic theory of matter, however, the paradoxical situation of homoeopathy defying the atomic theory of matter is going to remain.

Samuel Hahnemann, founder of Homoeopathy was a German physician, chemist and medical translator. He was disheartened by the harsh, often dangerous medical methods commonly used in his time. He sought to discover a more gentle and effective type of medicine. Hahnemann expanded the ancient idea of "like cures like" into a complete medical system for the first time. He coined the term "Homoeopathy", from the Greek roots for 'similar' and 'suffering', to describe the new system that he evolved from the law of similars.

Although many of Hahnemann's contemporaries attempted to discredit his ideas as being radical and contrary to accepted medical theory, homoeopathy was such a successful medical innovation that it spread throughout much of Europe and India in no time. The growing popularity of this virtually harmless system has continued to this day. It provides a safe, effective, natural and non-toxic treatment for many acute and chronic illnesses.

Satish Dhawan

(1920–2002)

Famous Scientist

Karma yogi is an odd description for a scientist. But Professor Satish Dhawan's life was an embodiment of such values. If Vikram Sarabhai was the father of India's space programme, Dhawan mothered it through its most critical years. Before he took over as chairman of the Indian Space Research Organisation (ISRO), Dhawan was a brilliant professor of aeronautical engineering at the Indian Institute of Science, Bangalore, becoming its youngest director when he was 42. Teaching was his first love, so when the late prime minister Indira Gandhi chose him to head ISRO in 1972, he agreed only on the condition that he continued as IISc. director. Dhawan gave ISRO its mission approach that set targets for every programme and rigorously evaluated them. During his tenure, India entered the exclusive space club by building its satellites and launching them. Always projecting his team, Dhawan camouflaged his humaneness with a no-nonsense approach. His twinkling eyes usually gave him away.

Satyendra Nath Bose

(1894–1974)

Bengal, at the turn of the last century, was an interesting place. English education, the growth of a new economy and other social and political forces were creating a new urban culture, and with it, a new enlightened middle class. Satyendra Nath Bose, born in 1894, belonged to the changing times.

S.N.Bose's grandfather had migrated to Calcutta from his village, a traditional centre of learning in the Nadia district of Bengal. British administration had thrown up new opportunities, prompting an exodus towards the metropolis. As an accountant working for the government, Bose's grandfather travelled throughout northern India. His son, Surendra Nath, an engineer in the railways, toured the northeastern region extensively. Education and faster modes of transport were bringing Indians together. They were now closer than they had ever been in a long and chequered history. Nationalism was a new force. A new spirit of self-help led to a demand for Indian enterprises and swadeshi goods. Surendra Nath Bose was one of the first Indians to set up a small chemical factory. Satyendra Nath, had therefore, two generations of English education and the tradition of a liberal outlook on life behind him.

Born in a family of six daughters, the young Satyendra Nath was naturally the focus of attention. His unusual gifts were recognized quite early and the teachers in his school doted on him. The school, founded by David Hare and Raja Rammohan Ray, had excellent teachers. They created in the boy a taste for literature and history which was to remain with him all his life. He was particularly indebted to the mathematics teacher who was very impressed with his extraordinary pupil. In one test, he gave Satyendra Nath 110 marks out of 100 because his pupil had also worked out all the alternatives in the question paper. In college, Bose was with a batch of bright students, including Meghnad Saha. Many of them were to achieve great success in later life. In the M.Sc. examinations, Bose came first in mathematics, with Saha coming a close second.

S.N. Bose was looking for an opening when both Saha and he were offered lecturerships by Sir Asutosh Mookerjee in the newly-opened College of Science of Calcutta University. Both of them joined the applied mathematics department, then switched over to physics. Sir Asutosh had played a key role in determining the shape of things at the Calcutta University. It was he who threw a challenge to students of mathematics with no formal background in physics, and assigned to them the difficult task of conducting physics courses at the postgraduate level. This was before 1919. The subjects Bose and his friend Saha had to teach included relativity, for which there were no books in English. And it was here that Bose's knowledge of German (he also knew French) came in handy. Along with Saha, he translated Einstein's papers on the theory of relativity from the German.

The war in Europe had put a stop to the import of new books and periodicals, but S.N. Bose read whatever he could lay his hands on. Sir Asutosh allowed the young lecturers to use his large personal library. Bose and Saha were thereby exposed to some of the latest ideas in physics and mathematics which they would not have obtained otherwise.

Bose produced research papers both singly and jointly with Saha. Then an offer came from Dhaka. When S.N. Bose joined as a reader, Dhaka University was still new. Many things needed to be organized. Most important for him was the setting up of a separate science library and getting a regular supply of journals. News of the latest developments in physics came to him via Calcutta. Then he came across a new book by Max Planck, one of the founders of modern physics. The book was a gift from D.M. Bose, a nephew of Jagdish Chandra Bose. The book presented new and exciting ideas to work on. Planck is best known for his quantum theory—a theory which upset the idea that energy was a continuous flow. On the contrary, Planck propounded that energy came out (or was absorbed) in discrete bundles or quanta. But Planck's work on the distribution of energy from a black body did not satisfy Bose. There were certain ad hoc assumptions which did not seem very logical. This set him thinking. Finally, he was able to hit upon a new solution. What he found was actually more than a formula—it was a new concept altogether.

He sent his paper to the philosophical *Magazine* and in a moment of daring, sent another copy to Einstein for his comment. Here is an excerpt from what Bose wrote to Einstein:

I have ventured to send you the accompanying article for your perusal and opinion. I am anxious to know what you think of it. You will see that I have tried to deduce the coefficient $\pi \dfrac{v^2}{c^3}$ in Planck's laws independent of the classical electrodynamics only assuming that the ultimate elementary regions in the Phase space had the content h^3. I do not know sufficient German to translate the paper. If you think the paper worth publication, I shall be grateful if you arrange for its publication in Zeitschrift fuer Physik. Though a complete stranger to you, I do not feel any hesitation in making such a request. Because we are all your pupils though profiting only by your teachings through your writings. I do not know whether you still remember that somebody from Calcutta asked your permission to translate your papers on Relativity in English. You acceded to the request. The book has since been published. I was the one who translated your paper *'Generalised Relativity'*.

The paper was published in German with a remark by Einstein, 'In my opinion, Bose's derivation of the Planck formula signifies an important development'. Bose's paper had an immediate as well as long-term impact on some fundamental problems of physics. Many are of the opinion that his work is by far the most important contribution to science made by an Indian, but he was not awarded a Nobel Prize. However, the concept—Bose Statistics—and the particles named after him will always remind us of his achievements.

In 1924, Bose went to Europe on study leave. His first halt was at Paris. There, Bose divided his time between Madame Curie's laboratory and that of the de Broglie brothers. Meanwhile he sent another paper to Einstein for comment. It was on 'Thermal Equilibrium in Radiation field in Presence of Matter.' The paper was duly translated and published but this time Einstein was more critical. He did not agree with some of the observations. However, when Bose went to Berlin he was warmly welcomed by the scientific community. Unfortunately, he could not do any sustained research work under Einstein, even though he did profit from a continuous exchange of ideas.

He returned to Dhaka after two years. After the well-equipped laboratories of Europe, Dhaka must have seemed hopeless but he was not disheartened easily. Nor did the prospect of working in the advanced centres of learning tempt him away from his mother-land where first-rate research was almost impossible, resources were limited and the government indifferent. He was determined to conduct the kind of experiments he had seen being done in Europe. He worked hard to organize a modern laboratory with facilities for research in X-ray spectroscopy, X-ray diffraction, magnetic properties of matter, wireless and so on. Dhaka slowly became an important centre for physics. Scientists like K.S. Krishnan and S.R. Khastagir went to work there.

S.N. Bose had a brilliant mind, which was equally at home in mathematics, chemistry, biology and physics. It is obvious that he could have reached greater heights of glory, had he so wished. The limited field of specialization, that of mathematical physics, did not satisfy his intellectual curiosity, and thus he never made a concentrated effort in one particular direction. Instead, he was soon guiding the research work of students belonging to other disciplines.

In 1944, he was elected as general president of the Indian Science Congress. In 1945, after a long stay in Dhaka, Bose moved to Calcutta to take up his appointment as the Khaira Professor of Physics. The following year all work came to a standstill because of communal riots. After Independence, however, changes were swift. The Saha Institute was founded; the Institute of Radio Physics and Electronics became a department of the university; the Khaira Laboratory achieved a considerable amount of work in X–ray crystallography. From 1948 to 1950, Bose was the President of the National Institute of Science. He visited Europe almost every year, either to take part in conferences or to present papers. In 1952, he was nominated a member of the Rajya Sabha, and he continued to be a member till 1958. In 1954, the nation honoured him with the title of Padma Vibhushan.

After retirement, Bose moved to the Visvabharati University at Shantiniketan as its vice-chancellor. He saw scope for much improvement there, particularly in the toning-up of science courses. He outlined a three-stage teaching programme and made plans for establishing a science institute. But opposition to his plans was growing and after a brief stay, he returned to Calcutta. Calcutta University made him an emeritus professor. He moved his office to the Indian Association for the Cultivation of Science after he became the National Professor. Mentally as alert as ever, he was nevertheless getting on in years. He died on 4 February, 1974, after an attack of bronchial pneumonia.

Bose had become a legend in his lifetime. People adored him, even though not everybody could fathom the mystery of his famous equation. People felt he was one of them. Perhaps no other scientist in our country enjoyed the kind of popularity that Bose did.

Bose loved knowledge for its own sake and was not a careerist. With his profound knowledge of the many branches of science, he often helped both students and colleagues by his valuable suggestions. He did not care for the publication of research papers. A doctoral thesis was not all-important to him. And so, even though Bose had a fine analytical mind and was clearly in the front rank of science, his total output in terms of published papers is small. His paper published in 1924 was hailed all over the scientific world as a brilliant piece of work. His next important publication appeared only in 1934. As a matter of fact, Bose was a most unconventional scientist. Basically a physicist and a mathematician, his mind explored a wider field. He worked with chemists, set up an organic chemistry laboratory at Dhaka and worked on the structure and stereo-chemistry of several alkaloids and other organic substances. In 1954, at the age of sixty, he published his researches on unified theory—another major contribution in mathematical physics.

Bose was deeply concerned with education. He believed that much of the waste in our system of teaching came from using a foreign language as a medium of instruction. From his interaction with hundreds of students, he concluded that a language, other than the mother tongue, often acts as a barrier between the student's comprehension and his expression. When Bose went to Japan to attend an international seminar, he found that in spite of the fact that most Japanese scientists understood English, all the deliberations were carried out in Japanese. In Japan, he was told, even technical education was imparted in Japanese. Bose was very impressed by this fact. It also strengthened his case. After coming back from Japan, the idea of a vernacular-based education obsessed him. He spoke endlessly on the subject—in public meetings, seminars, even during his convocation addresses. To show his critics that he was serious, he gave the Saha Memorial Lecture on an aspect of cosmology in Bengali.

He founded the Bangiya Vijnan Parishad with the sole objective of promoting and popularising science through the vernacular.

He was modest by nature and hardly ever talked about his own work. Only once did he make a lighthearted reference to it. The occasion was physicist Dirac's visit to Calcutta. Professor P.A.M. Dirac, his wife and S.N. Bose had got into a car. The professor and his wife were sitting at the back while Bose got in front with driver. Just then, Bose spotted some students and asked them to get in. Dirac protested mildly saying that perhaps there would not be enough room. Bose laughed and said, 'We believe in Bose Statistics. In Bose Statistics, things crowd together—which, of course, is an over-simplification.'

S.N. Bose received the honours due to him late in his life. He was made an F.R.S. as late as in 1958, and various national awards and honours were conferred on him only after Independence. He did not receive a Nobel Prize, justifiably his due, but he did win something far more precious—the affection of a vast majority of Indians, who were uninterested in scientific discoveries and loved S.N. Bose for what he was. And this, indisputably, was a well-deserved reward.

Boson, an elementary particle, is named after him.

Shanti Swarup Bhatnagar, Dr.

(1894–1955)

Shanti Swarup Bhatnagar Memorial Award is given every year for outstanding research by the Council of Scientific and Industrial Research (CSIR). It was instituted in 1958 in the honour of its first Director General Dr. Shanti Swarup Bhatnagar. The awards are given in the fields of Physical Sciences; Chemical Sciences; Biological Sciences; Earth, Atmosphere, Ocean, and Planetary Sciences; Engineering; Medical Sciences and Mathematics (alternate years). Each award carries a cash prize and a certificate.

Shanti Swarup Bhatnagar was a renowned Indian chemist. He was born on Feb.21, 1894, at Behra in the district of Shahpur in West Punjab. His father died when he was a baby. He was brought up by his maternal grandfather at Secunderabad and had his schooling in Lahore. He obtained his M.Sc. from the Punjab University in 1919. After taking his D.Sc. from London University under Prof. Donan, he worked under Prof. Haber at Kaiser Wilhelm Institute, Berlin, and later under Prof. Freundlich, an expert on colloids.

He was a Professor of Chemistry at Banaras Hindu University from 1921-24. From 1924 to 1940 he worked as Director of the University Chemical Laboratories, Lahore. There he made significant contributions in the field of physical chemistry, especially in magneto-chemical studies. He also wrote a book on magneto-chemistry.

He became the first Director General of Council of Scientific and Industrial Research (CSIR) in 1940 and held this post till his death. He was knighted in 1941 and won many academic and official laurels, both in his country and abroad. In 1943 he was made a Fellow of the Royal Society of London. In the same year the Secretary of the Royal Society Prof. A.V. Hill visited India to advise the government on the coordination of scientific research in India. Dr. Bhatnagar was one of the members in the meeting along with Hill, Saha and Bhabha.

He realised the importance of atomic minerals and initiated a nationwide search for ores. He convinced commercial companies to support research like Burmah Shell, Assam Oil company etc.

In 1946, when Pt. Nehru was the head of the Interim Government, Dr. Bhatnagar took up his views on the development of science in India to translate them into reality. He concentrated on applied sciences and managed to get substantial funds from industrialists for the building up of research laboratories. He opened a chain of National Research Laboratories in India.

When he died on January 1, 1955, Bhatnagar was holding a number of very important posts — secretary to the Ministry of Natural Resources and Scientific Research, director of CSIR, member of the Atomic Energy Commission and Chairman of the University Grants Commission.

This efficient and dynamic man of tremendous zeal, energy and foresight had many pleasant facets to his personality. He was the author of a volume of verse in Urdu and was given to scribbling humorous verses after important meetings, such as the one with Sir Winston Churchill. He is also remembered for his witty puns. On one occassion, when asked by an American student if he had any American pen-friends, Bhatnagar replied, 'I have quite a few real American friends, but the only pen-friend I have in America is my *Parker 51* and even that friend slips, out of my pocket now and then!'

Thanks to Bhatnagar's uncanny foresight and superb capacity for action, India is one of the leading industrial powerhouses of the world today.

Srinivasa Ramanujan

(1887–1920)

The Amazing Saga of A Great Mathematician

The story goes that when the well-known Indian cartoonist R.K. Laxman visited England for the first time, he was faced with a remark that the Indian contribution towards the world of mathematics was *zero*. Hearing this queer remark, he was flabbergasted, till he realized that it was a compliment rather than a sneer.

It is universally accepted that those were the Indian mathematicians who evolved the "zero" entity. The names of such mathematicians like Aryabhatt and Bhaskaracharya who contributed to the development and enrichment of mathematical thought is familiar to most of us.

In the nineteenth century particularly in its second and third decades, there was this rare type of a mathematical genius called Srinivasa Ramanujan. He amazed the educated community of this country and abroad by the ingenuity of his mind, extraordinary memory and number serendipity–the faculty of working unexpected and happy discovery of numbers by accident. He was deeply acquainted with the behavioural pattern of each number less than 10,000.

In the words of a psychoanalyst: "It is not possible to explain the phenomenon of Ramanujan except on the hypothesis of the ever-increasing purvajanma-vasana, the psychogenetic force gaining in momentum all through the march of a soul from embodiment to embodiment."

As further elaborated: "But there seems to be some supreme law of Nature by which such phenomenal men appear in this world from time to time as a result of some social and hereditary factors. Perhaps, such mysterious manifestations represent the immemorial wisdom of the East.

While employed as a clerk in the Accounts Office, Madras Port Trust, Ramanujan attracted the attention of prominent mathematicians. He used to present to Mathematical Societies in England a number of theorems mostly in pure mathematics pertaining to the theory of numbers and elliptic functions which he had proved. Many of these works were quite new and others had been anticipated by writers of whom he had never heard of and of whose works he was ignorant.

The story goes that one day Prof. Ross who was teaching Post Graduate class in mathematics at the Madras Christian College came to the class-room and showed his students Ramanujan's Quarterly Report which he had received from Cambridge the other day and said that a theorem contained in it had also appeared in a Polish periodical received in that day's post. Ramanujan did not know the Polish language and he 'divined' what the Polish Mathematician had thought.

Ramanujan showed his precocity in mathematics when he was yet a school boy. There are a few stories about his high mental mathematical genius. Soon after he had begun the study of Trignometry, he discovered for himself "Euler's Theorems" for the sine and cosine and was very disappointed when he found later, apparently from the second volume of Loney's Trignometry that they were known already.

One day his teacher in arithmetic said, "If three fruits are divided among three persons, each would get one. Even if 1,000 fruits are divided among 1,000 persons each would get one. From this he generalized that any number divided by itself is unity. This made Ramanujan slightly puzzled. He asked, "Is zero divided by zero is also unity?" The teacher had no answer.

Ramanujan joined College in 1904. He came across, through a friend of his, Shoobridge Carr's book '*A synopsis of elementary results in pure and applied mathematics*'. This influenced him immensely. It contained 6,000 theorems in various branches of mathematics.

Sometimes Ramanujan would embarrass his professors by his ingenuous queries. Once his Professor of Physiology asked him and his colleague to dissect a frog, whose structure he said was that of a man. When they had given chloroform to the frog with great abhorrence, Ramanujan asked the Professor "Sir, are the sea-frogs chosen for dissection because we are all well-frogs" (hinting thereby the insignificance of man as against other marvels of Nature). The Professor smiled and gave no reply. Ramanujan further asked whether there was the Serpent (i.e., the Serpent Power or Nidi in man—the Kundalini Sakti) in the frog. The Professor sidetracked the issue.

Ramanujan could not however do well in his career in the college due to his excessive devotion to Mathematics at the cost of other subjects. He even left his studies for some time. He could not qualify the F.A. Examination in which he appeared privately in December 1907. Subsequently he was married to a girl of nine named Janaki in 1909 and so it became necessary for him to find employment.

In 1912 when he joined the office of Port Trust of Madras, his first substantial paper had already been published a year before. It was however on his entry into the office that another paper of his was published and his exceptional ability was understood. He was awarded an overseas scholarship by the Madras Government and he went to England for research work in

Mathematics in 1914. He became a Fellow of Royal Society on 28 February 1918 and a Fellow of Trinity College, Cambridge, on 13 October the same year. He was the first Indian to be so elected. The fellowship carried an honorarium of £ 250 a year for six years.

During his four years' stay at Cambridge, he came nearer Prof. G.H. Hardy, who was impressed by his profound and invisible originality. In fact, Ramanujan was Hardy's discovery. He was the first 'competent' person to chance to see some of his works and at once recognised what a treasure he had found. To Hardy, though Ramanujan's pecularities were like other great men, he was a man in whose society one could take pleasure, with whom one could drink tea and discuss politics and Mathematics.

Ramanujan's health had betrayed him in England in the summer of 1917 because of which he had to return to India on 27 February 1919. He died on 26 April 1920 due to TB. He was hardly 33 at that time.

In his book on Ramanujan containing Twelve Lectures based on subjects on his life and work, Prof. Hardy pertinently points out that the death of Ramanujan was not so pathetic and painful (Abel and Galois also died very young) as the damage done to his intellect during period of 18 years to 25 years—critical years in a Mathematician's career due to the inelasticity in rules of promotion in the educational system. Had Ramanujan been encouraged in the educational field and financed a bit by a few friends, the world would have gained much more by this great Mathematician.

Ramanujan was born on 22 December, 1887 in a poor Brahmin family of Erode near Kumbakonam in Madras. His father was a clerk in a private firm in Kumbakonam. As Ramanujan's family was poor, it could not afford his further continuance in studies after his failure. Consequently he left his college. He had to bear a great pecuniary distress, but the desire in him to practise Mathematics never faded. He serves as a lesson to those who do not study for want of suitable and favourable circumstances in the house. Once a friend and his cousin came to his house to make him go for a walk. He was found under a cot solving his mathematical formulae. He adopted this novel way of duping his parents lest they should heckle him for not going out for some employment. There is another example. Sometimes he had no proper sheets of paper for his notings for want of money. He would collect bits of papers from here and there and write on them on both sides. When these were full, he re-wrote in red ink on the very sheets already written in blue ink.

As a man, Ramanujan was somewhat stout built but what attracted most in him were his shining eyes. The eyes, as we know, are the mirror through which a man's personality is reflected.

Ramanujan was a strict vegetarian. At Cambridge, he cooked his food himself. He felt great difficulty during the period of his illness, as he would eat food cooked only by the men whom he liked. As he was anaemic, he was advised to take two raw eggs daily but he would never agree to the proposal. His faith in vegetarianism never wavered at any stage of life.

As regards religion of Ramanujan, "He had definite religious views. He had a special veneration for Namakkal goddess. He believed in the existence of a Supreme being and in the attainment of God-head by man. He had settled convictions about the problem of life and after." He used to lecture on God, Zero and Infinity, a topic which was too intelligible to his colleagues.

Ramanujan's wife once reminiscenced how she used to collect the scattered papers of her husband on which he had some notings, during the period of his illness after his return from abroad. He was too weak to get up and keep the papers in an orderly manner.

Ramanujan still lives in the hearts of the people. Various memorials (portrait, Ramanujan Chair, Ramanujan Institute) have been made in his honour. The collected papers and his three notebooks (one of which was the 'lost' notebook) have been published. On his 75th birthday, a postal stamp was issued in his honour and the name of the Town Hall School Kumbakonam, was named after Ramanujan.

Ramanujan's birth centenary was celebrated in 1987 in U.K. and U.S.A. in the form of lectures, seminars and research papers.

Only the mathematicians can judge the work done by Ramanujan in such a short span of life. To an ordinary man he serves as an example of singular devotion to a cause in spite of heavy odds in life. He left behind a legacy which is to be remembered by the student community all over the world.

Very few people in India would be knowing that an American, Alle Selberg, born and educated in Norway took up Mathematics as a career just to do research work in Princeton University to work on Ramanujan's Collected Papers.

Ramanujan's ingenuity of mind and quick wittedness can be gauged by a single anecdote. When he was in a state of illness, a few months before he had returned to India, Prof. Hardy came to see him in the hospital. Ramanujan asked Hardy about the number of his taxi-cab. Hardy replied, "1729, not a particularly interesting number".

"What"?, replied Ramanujan, "Don't you know that it is the smallest number which can be expressed in two different ways as the sum of two cubes?" ($1729 = 9^3 + 10^3 = 1^3 + 12^3$).

Thomas Alva Edison

(1847–1931)

Great Genius

Slowly, noisily, the train puffed into the snowbound station. Most of those on board were friends and there was an air of expectancy from one rattling, sooty end to the other. Tom Edison's parties were always fun, always memorable (he usually produced some new and startling invention to entertain his guests) but an additional draw—and it had needed one to get all these people twenty-five miles out of New York on New Year's Eve—was that the party was actually being held in Edison's new laboratory, in the beauty spot of Menlo Park. Happy, laughing, expectant, the guests jumped down from the train and, by the smoky light of kerosene lamps, climbed into the horse-drawn buggies which awaited them.

Suddenly, the night and its moon, its few stars, disappeared, and the snow on the ground lit up like a million diamonds. The guests stared at each other, at the party clothes they hadn't really noticed before, and gasped.

They made their way up the short road from station to laboratory and now they could see the whole way lit by little incandescent pears that hung, hundred upon hundred of them, in a line from a wire just over their heads.

Outside the laboratory, Edison was laughing. "Like it?"

"Yes—yes, of course ! But what—what is it?"

"Electric light. Glad you like it."

The startled guests went on into the the brightly lit laboratory and the party began. In a few hours time it would be 1880, there was dancing and music and wine and light to be enjoyed.

Edison had just produced the world's first incandescent light. Half a century earlier Humphry Davy in England had made light electrically by passing a current through two sticks of charcoal. It had burnt dazzlingly, glaringly, painfully to the eyes, then burnt itself out in a few minutes. No one had been surprised : it was against the laws of nature for there to be "electric light"; it was a contradiction in terms, it was blasphemy, it was nonsense.

It needed only this to start Tom Edison in his search for a way of proving these opinions wrong. He was convinced it could be done, that he would do it . And yet he was modest: he knew only too well that what the public was now beginning to call his "genius", was "ninety-eight per cent perspiration, two per cent inspiration". He didn't mind being *"The Wizard of Menlo Park"*, that amused him—but "genius", no.

Although he undoubtedly was genius, Edison worked hard. He inherited his determination: his great-grandfather had been condemned to death as a British sympathizer in the War of Independence but had escaped to Nova Scotia; the next generation had migrated from Nova Scotia to Canada. Edison's father, Samuel, had been in a plot to overthrow the Canadian government and replace it by one more like that of the United States. His plot was discovered but he managed to escape—back to the United States. Through the help of a barge captain, Alva Bradley, Samuel was able to smuggle out his wife and six children, to Milan, Ohio, and it was here, on 11 February, 1847, that his seventh was born and named, in honour of the captain, Thomas Alva Edison.

When Thomas was seven, the family moved to Port Huron, Michigan, and it was there that Tom Edison went to his first and only school—for three months. After that—well, he was "addled", the teacher said, and, anyway, he was bottom of the class. His mother was forced to undertake his education and this, apart from the huge amounts he taught himself, ended at the age of twelve. He was going to be an inventor, inventors needed money for their experiments, and the only way he could get it was by getting a job. A new railway line had just opened from Port Huron to Detroit and here was his opportunity. His parents agreed to let him go: after all, his experiments about the house were a bit of nuisance; he'd sat on eggs to hatch them, given a playmate Seidlitz powders to blow him up like a balloon so he could fly, had even fed worms—suitably disguised—to the servant girl to make her fly, too.

The railway agreed to let him sell papers and sweets on board, without pay, and keep what profit he made. It seemed an ideal arrangement for a young inventor. The company allowed him to convert a part of the baggage car into a laboratory, and, as the train sat each day in Detroit for six and and a half hours, what time he didn't spend in his lab he devoted to studying chemistry and physics in the Detroit library. Then, each afternoon at four-thirty, he began the return trip, selling things, experimenting, selling more things, until he arrived home at Port Huron at ten.

Then he started, in the same baggage car, his own paper, printing it on a second-hand press. It was one sheet, the size of a handkerchief, and it always contained red-hot news which his telegraph operator friends passed to him before the stuff got into the country papers. His profits rose and he might well have concentrated on journalism had not an experiment with phosphorus set the baggage car on fire. Edison, his press and his lab., went out at the next stop.

Right after this disaster, he was able to save the small son of the Mount Clemens stationmaster from an oncoming train, and the man promised to make him "the best telegraph man in the country" and get him a job.

He kept his word and Edison did in fact become one of the best telegraphists in the country. Unfortunately, his first successful experiment got him the sack from his first job. He signed on in Stratford, Ontario, as operator, and on learning that he was expected to send the letter "A" each hour of the night to prove he was awake, he found it only too easy to make an attachment to the clock which would send the signal for him and let him sleep. But when the chief operator in Toronto decided to reply and got no response, he rode down to Stratford, saw the device and its sleeping inventor, and sacked Edison.

For months after this he was vagabond. Always untidy, he now became a scarecrow and succeeded in getting a job with Western Union Telegraph in Boston only by asserting that, despite his appearance, he was a better operator than their best. They arranged, tongue in cheek, for their best man, in New York, to send him a thousand words at several times the normal speed. The Boston operators crowded round for a laugh, but it was soon obvious that the New York man's speed was inferior to Edison's. Halfway through the message, seeing the looks on the operators' faces, realizing what they'd hoped for, Edison opened his key and tapped out a peremptory "Hurry" to his opponent. He was hired.

It was while he was with Western Union that Edison patented his first invention—a vote-recorder. It was a marvel of ingenuity, but nothing could have appealed less to the politicians of the day, so he set himself to producing other inventions more likely to meet with approval. He invented a duplex telegraph system in which two messages could be sent, in opposite directions, over the same wire, and while he perfected it he was simultaneously working on quadruplex telegraphy and a ticker-tape machine.

Bored with Boston, he moved to New York. He arrived, now twenty-two and quite penniless. For the first three days he found no one interested in him or his inventions. A friend arranged for him to sleep in the cellar of a company which ran a ticker-tape service over New York for stockbrokers, and it was his good fortune to have the system break down while he was on the premises. Messenger boys rushed in from subscribers, screaming for service, engineers tore from one installation to another. Amid the resulting confusion Edison quietly offered his services to the President of the company, who was almost tearing his hair in the centre of the room.

Edison had no trouble in finding the broken spring in the transmitter and within a minute the whole system was working. A grateful management made him foreman on the spot, with the comparatively huge salary of three hundred dollars a month. Edison was rich.

A year later, in 1870, when he was twenty-three, he resigned, having saved the money to open a workshop, and went into partnership with an electrical engineer, Franklin Pope. He began to manufacture his own ticker-tape machines in a workshop in Newark. Business was brisk, the shop needed little supervision, and Edison was able—by averaging only four hours' sleep a night—to devote more time to experiment. Between 1870 and 1876 he received patents for 122 inventions.

He married in 1871, and his devoted Mary must have seen very little of him, but they were idyllically happy. They had two children—nicknamed Dot and Dash—by the time his widowed father came to live with them in Newark, and the first thing Edison did with the old man was send him into the country to find a site for a bigger, better laboratory. Within a few days, Samuel Edison had found the village of Menlo Park in New Jersey. Here the young man set up his first big laboratory, which was, within a few years, to earn him an international reputation as "The Wizard of Menlo Park".

It was at Menlo that most of his major discoveries were made: his "phonograph" or gramophone, which was a logical development of his earlier morse-recording machine; the electric light, the electric tram; a practical telephone; to name only a few. In many ways Edison, who was a perfectionist, improved on the ideas of others, as in the telephone, for which he designed the carbon transmitter that increased the range and clarity of the instrument Bell had invented and made it unnecessary to keep bobbing the device up and down between mouth and ear. (Bell's electromagnet-and-diaphragm had been both transmitter and receiver, and had done neither job very efficiently.) But one discovery he left for others to develop: the "Edison Effect", the emission of electrons from a hot filament, led to the development by others of the thermionic valve on which were to depend radio telephony, the long-distance telephone, and talking pictures. With the thermionic valve small electrical currents could be amplified.

His phonograph performed faithfully for friends in 1877, repeating the rhyme, "Mary had a little lamb . . ." but so much work was required to make it into the instrument Edison dreamed of that he shelved it and went on to invent the light bulb. This, too, required months of work, even after the spectacular demonstration of New Year's Eve, 1879, before he was satisfied. He experimented with twelve hundred different varieties of bamboo before finding the ideal for the filament, and only then did he market the lamp.

In 1884 Mary died of typhoid. He was heartbroken but he dug himself yet deeper into his work, scarcely sleeping, pressing on with his schemes, opening a big plant in Schenectady. In 1885, as he watched the countryside go by from a moving train, he hit on the idea of moving pictures: at once he began drawings of his kinetograph and projector. He might at this point have worked himself to death, but for the timely arrival of the beautiful Mina Miller, eighteen years his junior, whom he married in 1886. He was then thirty-nine, with a name famous throughout the world. For Mina and the three children by Mary he built a large house in West Orange, and the next year, in order to be near them, he moved his laboratory there, from Menlo Park.

Mina bore him another three children and here she and Tom lived happily until his death, over forty years later. During much of this time he worked to perfect his moving pictures, but in the First World War he devoted his energies to synthesizing the chemicals which in peacetime had come from Germany. He designed a plant for making carbolic acid and was told it would take nine months to build: he went out and built it with his own hands and those of his chemical workers in seventeen days and produced seven hundred pounds of synthetic carbolic acid on the eighteenth.

Before his death at West Orange on 18 October, 1931—he was eighty-four—he had patented 1,300 inventions. Shortly before he died, American newspapers conducted polls to determine the ten greatest living Americans: *there was considerable difference of opinion over the lists, but each poll agreed that the greatest of all was Thomas Alva Edison.*

Vikram Sarabhai, Dr.

(1919–1971)

Father of Indian Space Research

Dr. Vikram Sarabhai was not only an imaginative and creative scientist but also a pioneering industrialist and an astute planner. He made significant contribution in the field of cosmic ray physics and in the development of nuclear power and space programmes. When Dr Bhabha died suddenly in 1966 in a plane crash, it seemed almost impossible to fill the vacuum but fortunately a worthy successor could be found in Dr Sarabhai. He took up the nuclear programmes with a challenge and also added fresh dimensions to the space research programmes in India.

Dr. Sarabhai was born on August 12, 1919 at Ahmedabad in a rich industrialist family. His early education was in a private school in Gujarat College at Ahmedabad. He then went to Cambridge, England, and obtained his tripos in 1939 from St. John's College. He then came back to India and started research work in the field of cosmic rays with Sir C.V. Raman at the Indian Institute of Science, Bangalore. In 1945 he went back to Cambridge to carry out further research on cosmic rays. There in 1947 he obtained a Ph.D. degree in the same field.

It was as early as 1942, when Dr. Sarabhai and his newly-married wife, Sreemati Mrinalini, were staying for some time in Poona that they conceived the idea of starting the Physical Research Laboratory in Ahmedabad. Soon after his return from Cambridge in 1947, Sarabhai started looking for a place for this project. He got a few rooms at the M.G. Science Institute to start the laboratory and Prof. K.K. Ramanathan was made its first director in 1948. The foundation stone of the new laboratory building was laid in February 1952 by Sir C.V. Raman and the laboratory was formally opened in April 1954. Dr Sarabhai made the Physical Research Laboratory virtually the cradle of the Indian Space Programme just like Tata Institute of Fundamental Research was one such centre for the Indian Atomic Energy Programme.

Dr Sarabhai not only encouraged science but also devoted a good deal of time to industry also. For over 15 years he nurtured a pharmaceutical industry and he was also a pioneer of the pharmaceutical industry in India.

The first institution that Sarabhai helped to build was the Ahmedabad Textile Industry's Research Association (ATIRA). In building ATIRA he helped to introduce the scientific method in a traditional industry. He was only 28 when he was asked to organise and build ATIRA. From 1949-1965 he remained an Honorary Director of ATIRA. In 1962 he helped to found the Indian Institute of Management at Ahmedabad. From 1962-1965 he remained as an Honorary Director of this Institute. Dr Sarabhai was mainly responsible for setting up of the Thumba rocket launching station. In 1966, after the death of Dr Bhabha, he became the Chairman of the Atomic Energy Commission and carried it to new heights. He gave special priority to the Cyclotron project at Kolkata, which is now an advanced centre for nuclear research and the best of its kind in Asia.

Today, the success of space programmes in India is largely owing to the groundwork prepared by him in this regard. Dr Homi Bhabha put India on the nuclear map of the world and Dr Sarabhai did it in the field of space. Due to his efforts India could launch its first satellite, Aryabbatta, just three-and-half years after his death.

As a result of his achievements Dr Sarabhai became a world renowned figure in the field of space research. He was given the Bhatnagar Memorial Award for Physics in 1962; Padma Bhushan in 1966 and was awarded Padma Vibhushan posthumously. He was elected the Vice-President and Chairman of the U.N. conference on peaceful uses of outer space in 1968. He presided over the fourteenth General Conference of the International Atomic Energy Agency.

Dr Sarabhai died suddenly on December 30, 1971, at the age of 52 when he was at the peak of his achievements. It was a great loss to India and the Indian science in particular. He shall always be remembered for bringing a new totality of approach to science.

Wilbur and Orville Wright

Conquerors of the Air

I

One mid-December day in the year 1903 two young men faced one another on the smooth, wind-swept sands of Kitty Hawk, North Carolina, USA. One had a coin balanced on his thumbnail and, when the other called, he flipped the coin into the air. When it came to rest at their feet Orville Wright looked at his brother and said:

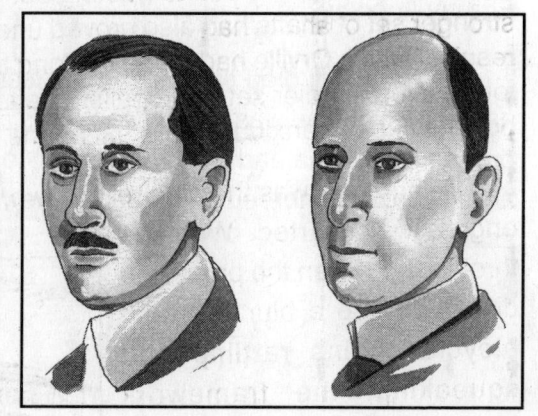

"Go ahead, Wilbur. You called right. You get first chance." If they were excited neither of them showed it; though what they hoped was going to happen in the next few minutes might make them both famous, and probably rich. Without another word they turned towards their dream child.

It was an odd-looking contraption of spars and frameworks covered with fabric. It resembled our ideas of an aeroplane only in that it had wings and an engine, plus two propellers. The engine had been made in the back room of the Wrights' cycle repair shop at Dayton, more than 500 miles away, even as the crow flies. The propellers, too, were home-made.

There were no landing wheels; the aeroplane being perched on a small trolley which, using materials from the shop, ran on two bicycle-wheel hubs. The trolley was designed to run along a launching platform, the total cost of which was four dollars, about $3.60 in English money today.

Comfort for the pilot had not been thought of. He was to lie on the lower of the two wings, and would manipulate not only the engine controls, but also the wing warping gear supposed to keep the plane on an even keel. The 12-horse-power engine would drive the two propellers by means of ordinary bicycle chains running around sprockets on the ends of the propeller shafts.

There had already been lots of trouble with these twin propeller shafts. A backfire when they first tried the engine and resulted in one of the shafts twisting out of true. A second and stronger set of shafts had also proved unequal to the strain as a result of which Orville had made the long train journey to Dayton for an even sturdier set of propeller shafts. Now, and they must both have been anxious, the great moment had come.

Wilbur, who was the elder of the two, got into position. The engine was started. With the throttle wide open the propellers dissolved into a blur and with bicycle chains rattling and squeaking, the framework creaking and vibrating, it was obvious that something must happen soon. Either the plane would move or drop to pieces.

Orville bent down and slipped the cable which anchored the plane, and at once it began to move forward, its speed increasing quite rapidly. The handful of spectators, some on the nearby Kill Devil Hill, ceased talking and watched. Not one of them really believed this crazy thing would fly, and within seconds their fears were realised. They started off at a run down to the level sands, hoping that Wilbur Wright had escaped serious injury. The plane had seemed as if it was about to lift nearing the end of the launching platform, and then without warning it had tipped forward and driven its front elevator into the sand. The roar of the engine stopped.

Wilbur got up and shook himself, then looked at Orville. The front elevator was smashed, so there could be no hope of a further try that day. Harry Moore, a cub reporter from the *Virginia pilot*, desperately keen to get an exciting story, soberly put away his pencil and notebook, then asked:

"What do you do now, Mr. Wright? My editor was kind of expecting me to . . ."

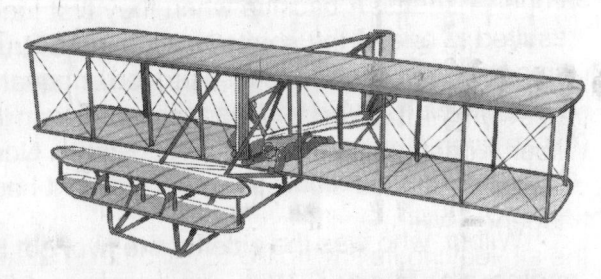

"Get a good story, eh?" Orville said, and made a little clicking sound of annoyance. "Well, mebbe you will. Come back on the 17th. Yeah, I guess we'll have it right for that date. We'll give you a story . . . a scoop of a story."

Wilbur dusted the sand from his trousers, straightened his tie, and turned his cap the right way round. The days of flying kit had not come, and both the Wrights were in everyday clothes. Wilbur had done one thing only to make sure he was dressed for such an epoch-making event as the first powered flight into the air—he had turned his cap back to front so that the wind would not get under the peak and blow the cap off.

II

For two more days the sands at Kitty Hawk were left to the gulls and the winds blowing in from the Atlantic; then the Wrights appeared once more with the damaged plane repaired.

They had spread the news about that they were going to fly on December 17, but only a handful of spectators had arrived to see what these 'crazy Wright boys' would do. There was much more excitement at the sea's edge, for one of the United States's first submarines, the *Moccasin*, had broken away from the tug which was towing it, and had gone hard aground, abreast the Currituck Light.

Alpheas Drinkwater, at that time a telegraphist for the United States Weather Bureau, had been invited to come; for if there was any great news, he would be the one to flash it out to the world. Alpheas did not come. He was too busy keeping the U.S. Navy Yard people informed about the efforts of salvage tugs to get the *Moccasin* off the beach.

For the second attempt Orville Wright laid himself on the lower wing. Again the engine was started, and again the whole framework of that first plane rattled and shuddered. This time Wilbur Wright released the anchoring cable. Once more the frail contraption moved along on its trolley, and then the incredible happened. With Orville Wright in command the plane lifted into the air, leaving its tiny trolley to roll off the runway and kick up a little spurt of sand.

Orville felt the sudden smoothness, realized that his front elevator was coaxing him into the air far too quickly, and tried to adjust it to give level flight. He was already ten feet off the ground.

Instead of straightening out into level flight the plane nosedived and, 12 seconds after taking off, the first flight had ended. In that short time the first powered plane to fly had covered about 100 feet.

There were three more flights that day, and the best of these saw the aeroplane stay aloft for 59 seconds and cover 852 feet. *It was the birth of the flying machine.*

Alpheas Drinkwater, still keeping the U.S. Navy Yard informed of the work going on about the stranded submarine *Moccasin*, was irritated that evening when, because of the failure of another telegraph wire, he had to send off several telegrams, one of which was to Katherine Wright, and read:

"Flight successful. Will be home for Christmas."

That message should have shaken the world, for it meant the dawn of the era of flying; but it passed unnoticed. Young Moore, the cub reporter from the *Virginia Pilot*, let himself go with a report which told how, Wibur Wright had soared into the air with bird-like ease, and flown three miles over the sea. His exaggeration of the accomplishments of the Wright brothers did not bring him renown. Editors simply refused to believe it. One or two papers printed paragraphs, and it is said that the *British Daily* Mail inserted a tiny paragraph at the bottom of a page headed: "*Balloonless Airship*".

III

How did all this come about?

Perhaps the editors should not be blamed too much for refusing to believe that man had conquered the air. Other attempts to fly had ended in disaster. Otto Lilienthal, a German, had built a bird-like glider with which he made over 2000 flights before crashing to his death. In the year this German pioneer died, an American, Dr. S. P. Langley, secretary of the Smithsonian Institute, built steam-powered model aeroplane. It flew extremely well, covering more than half a mile at the amazing speed, for those days, of 25 miles per hour.

Given 50,000 dollars by the U.S. Army towards the cost of building a man-carrying aeroplane, Dr. Langley built one with a specially designed petrol engine of 52 horse-power. It was to be launched from a houseboat moored in the Potomac river; but something went wrong with the launching apparatus and the machine simply ran along the runway and toppled from there into the river. Writers in the American newspapers made great fun of this, and for the time being attempts by man to fly were laughed at.

About this time Wilbur Wright was taking a great deal of interest in the efforts of the German, Otto Lilienthal, to conquer the air. Lilienthal had built his wings but had at length crashed to his death. Wilbur had not had an easy life up to then. Just when he was about to enter college he had an accident which disabled him for seven or eight years, during which time he had to stay at home—a period he spent in caring for his invalid mother.

In his spare time he watched the buzzards as they wheeled about in the air, and tried to discover the reason why they could circle so easily with what appeared to be the merest flick of a wing tip.

In 1890 Wilbur joined his younger brother Orville, who was then publishing a small, weekly newspaper, and began to get even more interested in flying. Orville caught his enthusiasm when Wilbur explained that he felt the secret of successful flight lay in having wing tips which could move as the wing pinions of a bird did.

Leaving the publishing business they went into a cycle repair shop. Here, with plenty of tools at hand, they began to construct their first glider. When completed it weighed a mere 50 lbs. It had a small wing span, but had something other experimental gliders had not possessed—a balancer at the front of the main wings (today we know them as elevators, and they are incorporated in the wings themselves). In addition to the balancer the two brothers fixed wires to the wing tips, so that they could be twisted in the same manner as the buzzard lifts or turns down the tips of his wing pinions.

Their first glider cost $4 and there was no place for the pilot to sit. He lay between the wings and worked the crude controls from that position. Now came their first visit to Kitty Hawk, where the winds were steady, the sands smooth and soft, and perhaps most important of all, there was a small hill off which the glider could be launched.

Not very satisfied with the glider they made many smaller models, and to see how they reacted to winds, fastened them on their cycle handlebars. Riding as fast as possible they would study the effect of the wind on them. Roads, however, were not like the smooth roads of today, and looking out for bumps and studying a model glider on the handlebars was unsatisfactory.

IV

One evening Wilbur startled his younger brother with a suggestion.

"I think we ought to be able to study the effects of wind with the models placed in a sort of tunnel."

"But where is the wind to come from?" Orville asked. "After all, you'd need something like a hurricane blowing through your tunnel to——"

"I've thought of that," Wilbur interrupted. "Suppose we built a small tunnel and put a fan at one end, the model at the other. We could blow wind through, and be able to study the effects much better."

Today's mighty wind-tunnels, where not only model aeroplanes are tested but also giant airliner models, were born when the first crude tunnel was made at Dayton by the Wright brothers. That small tunnel enabled them to build a third glider and solve a difficulty which had up to then seemed insuperable. It gave them the idea of fitting a rudder at the stern of the glider, and when this was tried out, with elevator and rudder linked by wires, they discovered they could manoeuvre their glider easily and with great smoothness. They could climb, bank, dive, turn, and keep a fair amount of control all the time. That year, which was 1902, they made nearly 1000 glider flights; one of which was record.

We think of record-breaking flights today in terms of round-the-world flights, without touching down, and being re-fuelled in mid-air by tanker aircraft; or flying across the Atlantic and back, breakfast at home, dinner in the evening at home, with a double crossing of the *'herring pond'* in between. A record flight in 1902 meant the Wright's glider had risen off the ground and not touched it again for a full 200 yards!

V

In the winter of 1902 Orville brought up the idea of fitting an engine to the glider. Wilbur, four years older than his brother, shrugged and looked very thoughtful.

"It's a question of weight, Orville. To carry the weight of an engine, petrol, and a man, the plane will have to be made bigger, and its probable weight will be in the region of five hundred pounds. I've been working it out, too."

"Sure, I know," Orville agreed; "but we can do it. Look at Langley; how big his machine is."

Wilbur nodded and smiled.

"Yes, look at it," he agreed, "but it hasn't flown yet. And in any case, where can we get an engine? Langley has got that brilliant engineer, Manly, to make him a special engine. He won't make one for us, even if we could afford it."

Orville sat staring at the big stove, glowing red; then his thin lips parted in a smile, and there was a twinkle in his eyes as he looked up saying :

"What about making our own engine?"

They sat in silence for several minutes, looked up, and nodded.

Throughout that winter they worked steadily and finally built a 12 horse-power engine. It worked but was not as efficient as Langley's 52 horse-power model. Now they faced the task of getting a propeller. Enquiries from ship builders convinced them that a propeller made like a ship's propeller would not do.

They finally designed a propeller with blades like small curved wings, and this proved amazingly efficient. To make sure they would get enough 'pull' to lift their plane into the air they decided to use two propellers, and drive these from their one small engine by means of ordinary bicycle chains. Later that year, as described earlier, they were the first men to fly in a power-driven aeroplane, going not at the mercy of the wind, but urged on by the might of a 12 horse-power engine. Their best flight that year lasted 59 seconds, and they attained what in those days was the astounding speed of nearly 30 miles per hour.

VI

Though they continued with their experiments throughout 1904 few people would believe they had really flown. The renowned American physicist, Simon Newcomb, had stated definitely that it was quite impossible for anything heavier than air to fly. Newcomb was a clever man, and people preferred to believe him rather than a handful of country folk from Kitty Hawk who continued to swear they had seen the Wright boys flying in the air.

In 1905 the brothers conquered one of their earlier troubles, a tendency to get into a tailspin when doing short turns, and in October of that year Wilbur made the longest sustained flight ever, keeping airborne for 38 minutes, flying round and round over a small circular course.

Efforts to interest the U.S. War Department were, however, without success. The 50,000 dollars the department had handed over to Langley, only to have him laughed at by every newspaper in America, seemed to have convinced the authorities that, as Newcomb insisted, flying was impossible for heavier-than-air machines.

Still thought of as 'those crazy Wright boys', Wilbur and Orville went back to Kitty Hawk sands in 1908 with a new machine, and this time they got Mr. D. B. Salley, a renowned free-lance journalist, interested. He saw them fly, and was convinced the world was on the verge of a new era, an age when man would fly as easily as a bird, but probably faster, and farther. He wired his story to leading newspapers, and none would use it. The *New York Herald* did send its own star reporter to Kitty Hawk to see what was going on. When he sent in a long story of a man flying through the air on a machine, swooping and diving, turning this way and that, then coming safely to the ground, the editor not only refused to publish the story, but suspended the reporter.

Oddly enough the man who set the seal of truth on these strange stories of men flying like birds, was the one-time telegraphist who had been at Curritucket Inlet for the U.S. Weather Bureau when the Wrights made their first flight five years earlier. Then he had refused to waste his time over such crazy antics as men trying to fly.

Now, however, he went on to Kill Devil Hill and watched what he described later as a 'two-winged contraption of split fir and cloth bear a man and a clattering engine into the blue'. He went home and swore out an affidavit which did two things, it put the *New York Herald's* star reporter back in his job, and gave the world the story of man's conquest of the air.

In 1909 Wilbur convinced even the U.S. War Department. Taking the machine to Fort Meyer in Virginia the brothers proved beyond doubt that they could fly, that they could make their machine do what they wanted, and the aeroplane was finally accepted as something which might possibly be useful to the United States if ever they went to war.

From that demonstration Wilbur went to Europe, and at Le Mans and Pau gave further demonstrations, for now the French were interested. King Edward of Great Britain went to see Wilbur fly. The barriers had been broken down, and the crazy Wright boys were now hailed for what they were, the men who had first conquered the air with powered machine.

Today the jet aircraft boom their way through the sound barrier; distances have been almost annihilated. A man can go to sleep in the Old World and wake up in the New.

VII

How strange that in the 1870's, according to a story published by the *New York Times*, Bishop Wright had a slight argument with the head of a college. The bishop had said that in his opinion everything that could be invented had been invented. The principal of the college shook his head slowly before saying:

"In fifty years' time men will learn to fly like birds."

Shocked, the bishop replied:"Flight is reserved for angels," never dreaming that the two small sons he had left at home, Wilbur and Orville, would be the first of the earth-born **'angels'**.

Wilhelm Roentgen

(1845–1923)

On a chilly day in 1895 the German scientist Wilhelm Roentgen was carrying out an experiment with the cathode ray discharge tube in his laboratory. The cathode ray discharge tube was completely covered by paper board so that light might not escape from any point. He felt that a piece of paper, which was lying a few feet away from the discharge tube and on which was coated berium bletino cyanide, was shining in the dark. This incident of fluroescence was also noted earlier by J.J. Thomson and many other eminent physicists but they did not feel it important enough for a detailed study. As they were ignorant of the nature of these rays so they named them as X-rays. Roentgen, however, took this accidental discovery very seriously, and for the next six weeks tried to find out all the properties of X-rays. And for that he ate and slept in his laboratory.

The German physicist Wilhelm Roentgen. He was awarded the Nobel Prize in 1901 for discovering X-ray. The X-rays were discovered accidentally when in 1895, he was carrying out an experiment. The discovery ushered in a revolution in medicine and physical science.

These new rays seemed to be emitted by the anode of the discharge tube and differed from the cathode rays in the fact that they could penetrate up to 2-metre in the air outside the tube, whereas the cathode rays were confined to the discharge tube only. When Roentgen discovered the opaque nature of other substances to X-rays he found that these rays could pierce a one-millimetre thick piece of lead. When he was putting a small disc of lead in the course of X-rays, he observed that the X-rays beside creating the image of the lead disc also showed the outer line of his own thumb. The varied opaque of flesh and bone being different, he could see the image of the finger-bones darker than the shadow. Hence, Roentgen concluded that X-rays could penetrate even those substances, which even light did not penetrate.

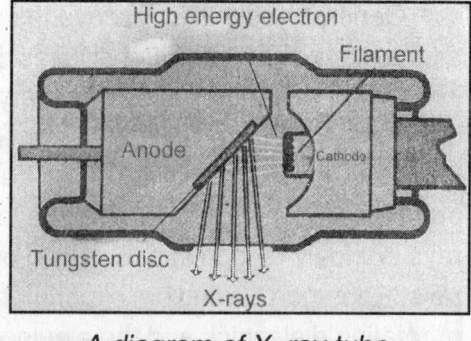

A diagram of X–ray tube

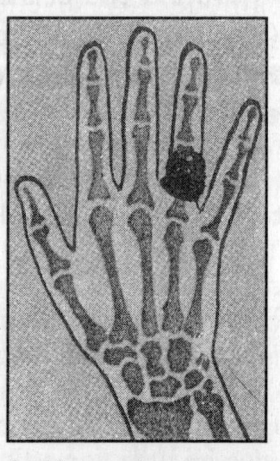

Roentgen made a number of accurate observations about these rays. For example, according to him, the wavelength of these rays should be shorter than the light-rays like electromagnetic radiation because these rays remain unaffected from the magnetic field. The results of his six-week long concentrated study were published in many leading journals. Roentgen himself called these rays as X-rays. He did not like anybody calling these after his own name. He made only one public speech about this new discovery. He declined to accept numerous honours for his remarkable achievement.

An X–ray photograph of the hand wearing a ring

Generally scientific discoveries become practically useful after a many-year-long extensive research. But X-rays were used just two months after their discovery when they were used to diagnose a fracture and its treatment in the New Hampshire Hospital. Keeping this in view, he was honoured with the prestigious Nobel prize for physics in 1901. The utility of these rays is widely recognised the world over. Beside discovering X-rays, Roentgen carried out experiments on the magnetic effects of rotating dielectrics and the electric occurence in crystals.

At the end of the 19th century, Roentgen left Wurzburg (Bavaria) for Munich to take up an assignment in physics there. He led a lonely life in Munich, and died there in 1923 at the ripe age of 77.

William Caxton

(1422–1491)

Not by virtue of the craft he fathered, nor for the practical success of the enterprise he boldly began, is the name of William Caxton honoured by his countrymen. He was a great and successful merchant who carried and upheld the good name of England abroad, and he was the man who brought the new art of printing from the Continent into England in the fifteenth century. But he was also a patriotic, pious and cultured Englishman, and he printed by the same ideals as those by which he lived. He was not only England's first printer, but one of her first educators, a moulder of taste, and an establisher of literary tradition.

With a national genius for giving people misleading tags, the British have, dubbed Caxton " the first English printer, " and let him go at that. The Englishman who loved his native language, and did more for it than all its patrons and hundreds of its writers, has been lost in the printer. Caxton was no mere introducer of a new trade; he was a pioneer and a creator, man of letters and an educationist who strove all his life for the benefit of his country's culture as expressed in its literature. Engand owes Caxton a debt of gratitude as her first printer, it is true, but she owes him a thousand times more for the spirit in which he printed. As a craftsman he was not exceptionally good, being far inferior to some of the Continental printers. " But... Caxton had noble conception of his duty as a printer," writes Mr. H. R. Plomer. " To him the mere mechanical process was a secondary consideration. His aim was to perpetuate such books, for preference those of English authors, or translators, who had the highest moral or literary value, and so long as the paper was strong and durable, type clear and readable, and the press-work correct, nothing else mattered. The road he set out to make was none other than to educate and brighten the lives of his countrymen by circulating hundreds of copies of the best literature at a low price where only half a dozen had been obtainable before, and those only by the rich."

William Caxton was born in Kent about the year 1422. The exact date of his birth is unknown, nor is there any record of his early years and education until on June 24, 1438, his name was entered in the books of the Mercers' Company as an apprentice to Robert Large, mercer, of London.

That fact gives us two others: that he was born about 1422-4, for he would hardly be less than fourteen nor more than sixteen when he was apprenticed; and that his family was influential, for the Mercers' Company was one of the most exclusive and powerful of the London Guilds, and Robert Large was an important member of it; he was elected Lord Mayor of London the year after Caxton had joined him.

For three years, Caxton lived with Large in Old Jewry, working and playing and leading the normal life of an apprentice. Then, on April 24, 1441, Robert Large died, leaving Caxton twenty marks* in his will. (worth about £300 in present-day money)

Presumably Caxton was transferred to a new master, and certainly in the same year he left England for the Continent. In 1471 he wrote that he had "continued by the space of xxx yere, for the most parte in the contres of Braband, Flandres, Holand, and Zeland."

We have few glimpses of the man during his rise to wealth and consequence. We know that he was pious and industrious, and he probably applied to his trade the enthusiasm he later brought to publishing. Within nine years he was a man of substance and credit, for in 1450 it is recorded that he and another mercer were sued for a sum of £110, equal to about £3,000 today.

Three years later Caxton visited London, and was admitted to the Livery of the Mercers' Company. He returned to Bruges and worked steadily, becoming eminent among the merchants of that town.

In 1462 Edward IV renewed the charter of the Company of Merchant Adventurers, a powerful guild that looked after the interests of English merchants trading abroad, and he gave permission to the Company to appoint a Governor at Bruges. A year later Caxton was fulfilling the duties of the governor, and soon he was officially appointed Governor of the English Nation at Bruges. His duties consisted of adjudicating in disputes among the merchants, supervising the importation and exportation of goods, and protecting the commercial interests of England.

In 1464 he took part in negotiations with Philip the Good, Duke of Burgundy, for the renewal of the trade treaty between England and Flanders. But the treaty was allowed to lapse and the English merchants in Bruges sought to overcome the difficulties by smuggling. Caxton received a letter from the Earl of Warwick telling him to stop this illegal trading. So it is evident that Caxton was kept busy with public business as well as his own trading. Nevertheless he found time to travel, and more important, to read.

The hostility between Flanders and England ended with the death of Philip in 1467. A year later his successor, Charles the Bold, marrried Edward IV's sister Margaret at Bruges, and in 1469 Caxton resigned his position as Governor of the English Nation and entered the service of the new Duchess of Burgundy.

A little while before this important change Caxton read a French manuscript by Raoul le Fevre, *Le Recueil des Histoires de Troyes*. The book gave him " great pleasyr and delyte," and he had the immediate inspiration to translate it into English. He began straightway, but after working hard through a score or so of pages became weary and disheartened and put the unfinished work away.

But his new patroness, the Duchess, became an inspiration and patron when she heard what he had been doing. She asked to see the work that had been done.

"And whan she had seen hem, anone she fonde defaute in myn Englisshe whiche she commanded me to amende and more ouer commanded me straytly to continue and make an ende of the resydue than not translated, whos dredfull comandement y durst in no wyse disobey becase y am a servant unto her sayd grace and resseive of her yerly fee and other many goode and grete benefets, and also hope many moo to resseyve of her highnes, but forthwith wente and labouryde in the said translacion aftyr my simple and poor conning also nygh as I can followyng myn auctour. ..."

The translation which he had begun at Bruges he now continued at Ghent, and eventually finished at Cologne. Then at Cologne, in 1472, he learned the new and wonderful craft of printing. Two years later he set up a press at Bruges with Colard Mansion, who was a calligrapher and illuminator of manuscripts, and from that press issued the first book ever printed in the English language—his own translation of the *Recuyel of the Historyes of Troy*. In the Prologue he wrote :

"And for as moche as in the wrytyng of the same my penne is worn, myn hand wery and not stedfast, myn eyen dimed with ouermoche lokyng on the whit paper, and my corage not so prone and redy to laboure as hit hath been, and that age creepeth on me dayly and feebleth all the bodye, and also because I have promysed to dyuerce gentilmen and to my frendes to addresse to hem as hastely as I myght this said book. Therefore I have practysed and lerned at my grete charge and dispense to ordeyne this said book in printe after the maner and forme as ye may here see".

Caxton worked with Colard Mansion for two years. In 1476 they printed *The Game and Play of Chess Moralised* which Caxton himself translated from the French. Thus the first two books printed in the English language were not the fruits of chance, but were deliberately chosen by their scholarly printer as being worthy to entertain and elevate his fellow countrymen. Caxton realized that printing had opened the golden gates of literature to all, and he was determined to produce work worthy of that wonderful opportunity.

Before the end of that year Caxton had left Mansion and the press at Bruges, and had set up at the sign of the Red Pale at the Almonry in the precincts of the Abbey of Westminster in London.

He began work at once, printing "small storyes and pamflets" and preparing to produce larger works. The first book which bears a date was the *Dictes or Sayengis of the Philosophres,* published at Westminster in 1477. The translation was made from the French by Earl Rivers, who asked Caxton to revise it before printing.

Earl Rivers became one of Caxton's constant patrons; among the most notable of these were Margaret, Countess of Richmond, Henry Bourchier, Earl of Essex, and William Fitzallan, Earl of Arundel, while many rich city merchants were his friends and helpers. Above all, Edward IV and his successors, Richard III and Henry VII, gave Caxton patronage and encouragement.

On June 15, 1479, Edward IV gave Caxton £20 for "certain causes and matters performed." To Richard III was dedicated the *Order of Chivalry,* while Henry VII himself commanded Caxton to translate and print *The Faytte of Arms.*

Soon after his arrival at Westminster, Caxton gave another proof of his desire to give Englishmen the best of English literature. By 1478 he had printed for the first time Chaucer's *Canterbury Tales* – a heavy task for his small press. He was in such a hurry to get this great work out that he chose a bad manuscript and issued a corrupt text. When he discovered this he was greatly distressed but some years later he procured a better manuscript from which he printed a second edition.

And now books flowed fast from the tiny printing press. Caxton's energy was amazing. He laboured at editing and translating, probably leaving most of the presswork to his foreman, Wynkyn de Worde, who carried on the shop after Caxton's death.

While he worked at the books he loved and at those works which he thought were fittest to publish for the general good, he also turned out Church service books and psalters. In connection with one Church service guide, Caxton published the first broadside and book advertisement to appear in England :

"If it plese ony man spiritual or temporal to bye ony pyes of two and three comemoracions of salisburie use, enprynted after the forme of this present lettre, whiche ben wel and truly correct, late hym come to westmonester in to the almonetrye at the reed pale and he shal have them good chepe."

Caxton's success in England soon attracted others to the trade. Oxford had a printer by 1478, and a year later another appeared in St. Albans. Then came a close rival to Caxton in John Lettou, a Lithuanian, who set up a press in the City of London.

Lettou's work was technically superior to Caxton's, and the competition led the Englishman to make improvements in his types and formats. In 1479 he began to use a new type, the Black Letter, and also the type known as Caxton No.4. He also resorted to the use of signatures and spaced his lines out to even lengths. In 1481 he published the *Mirrour of the World,* the first of his books to be illustrated. The illustrations were woodcuts and were of poor quality. None of Caxton's books were printed with title pages; some had no punctuation at all; some were only punctuated with the long comma, which consisted of a straight thin line between the words; some were punctuated with full stops and colons only; and he never numbered the pages.

But Caxton was always more interested in the matter than in the appearance of his productions. Piety, moral value, and literary worth were the chief criteria he employed in choosing books to print; but he did not neglect any work which would expound knowledge of any kind, nor did he despise good stories. Some idea of his tireless energy may be gathered from the fact that his tiny press at Westminster turned out over eighty books, of which twenty-one were translations made by Caxton himself! That his judgment and taste were popular is proved by the fact that many of the volumes he issued ran into two and three editions.

Besides those books already mentioned, these are the most interesting of Caxton's publications: *The Chronicles of England,* which was a work called the *Chronicles of Brute* brought up to within a few years of the date of publication (1480) by Caxton; this came to be known as *Caxton's Chronicle.* Cicero's essays on *Old Age and Friendship* in English together with the *Declamation of Noblesse in one volume (1481).* His own translation of *Reynard the Fox,* a clever German satire. Higden's *Polychronicon,* brought up to the accession of Edward IV by Caxton (1482). The *Golden Legend,* one of the most important and influential books printed in England at that time. Chaucer's *Hous of Fame* and *Troilus and Cressida, Æsop's Fables,* and Malory's *Morte d'Arthur.* He also published the works of the English poets, Gower and Lydgate. That short list is evidence enough of Caxton's taste and care for literature.

In 1491 Caxton was engaged on translating the *Lives of the Fathers.* At last he reached the end of the long task and concluded his version with these words :

"God thenne give us his grace to find in us such an house that it may please him to lodge therein to the end that in this world he keep us from adversity spiritual. And in the end of our days he bring us with him in to his Royame of heaven for to be partyners of the glorye eternal, the which grant to us the holy trynyte. Amen."

With that he laid down his pen—and his life; for Wynkyn de Worde tells us that his translation of the *Vitae Patrum* was finished by Caxton "at the last daye of his lyff."

He died in harness, with a prayer on his pen. And all who love literature should pray for William Caxton, not for the mechanical invention he brought to England, but for the ideals which he set steadfastly before him.

William Harvey

(1578–1657)

There are some scientific phenomenon with which we are too familiar, that it seems impossible that their truth could ever be doubted. It appears incredible that only a few centuries divide us from the discoveries of the principle of gravity, of solar motion ... and of the circulation of the blood. When William Harvey put forward his theory, or rather his observation, of the circulation of blood in the body, he shattered the medical beliefs of a thousand years and opened a new era of medical science.

Of how many books can it be said that their publication directly affected the personal well-being of every person alive or yet to be born?

Many books have been written which have altered the outlook of millions, changed social institutions, and even deflected the course, of history, but of few can it be said that their contents concerned the health not only of contemporaries but of all posterity, that they expounded a discovery vital to the whole of mankind. Yet such a claim may justly be made for a book published three centuries ago. At that time only a few read it and not all believed it. Today still fewer people read it, but none would dream of doubting it.

The book was William Harvey's work on blood circulation, published in 1628 in Latin under the title of *Exercitatio Anatomica de Motu Cordis et Sanguinis in Animalibus*. It overthrew the accepted doctrine of fifteen centuries. It proved that what men had thought about the movement of the blood in the human body was utterly wrong.

William Harvey, the man who discovered the circulation of the blood—a discovery that has been described as "an almost faultless solution of the most fundamental and most difficult problem in physiology" — was born at Folkestone on April 1, 1578.

His father was a prosperous merchant and William was sent to King's School, Canterbury, and afterwards to Cambridge. Whilst his work at the university was good, it was not sufficiently distinguished for the most prophetic of professors to suggest that the hard-working undergraduate was likely to gain immortality as the elucidator of one of the most important truths known to medical science.

In 1597 Harvey travelled to Padua, at that time the most famous school of medicine in the world. There, in the candlelit lecture hall of the university, he listened to Fabricius of Aquapendente, the great anatomist. From him he learned something that was later to prove the starting point of his great discovery. Fabricius had found that the veins in the human body had valves, and while he imparted this information to his students he was altogether unable to offer a suggestion as to the function of the organs. It remained for Harvey to realize that the valves prevented the flow of blood from any direction except towards the heart.

After becoming a Doctor of Medicine at Padua he returned to Cambridge, where he took a similar degree, after which he set up practice in London. It was about this time that he fell in love with Elizabeth Browne, the daughter of Launcelot Browne, who had been physician to Queen Elizabeth. The marriage proved of great use in helping him to gain initial entry to Court.

In 1609 Harvey was appointed physician at St. Bartholomew's Hospital. His task was to "attend the hospital one day in the week at the least through the year, or oftener as neede shall requye." He was to "give the poore full benefit of his knowledge," and to prescribe only medicines that would do the poor good " without regard to the profit of the apothecary." This last was a reference to the practice of the apothecaries, the chemists of the day, who used to buy up doctor's prescriptions to sell to their customers. Harvey apparently was not regarded as much of a doctor by the apothecaries, for none of them would give more than threepence for his prescriptions.

At the hospital Harvey would sit at a table, his patients seated before him on a settle. The physicians did not attend to patients in the wards, that being the surgeons' duty. Harvey would go into the wards only to treat a patient who was unable to walk to him.

His experience at the hospital was of the utmost value. He was able to collect a mass of data from the cases he attended and some of it made him ponder on what was soon to become the main the idea in his mind. For it had begun to dawn upon him that the age long theories on the subject of the movement of the blood in the human body were unreliable. He did not immediately realize where the errors lay, but that there were errors, and fundamental ones, he was becoming more and more convinced.

To Harvey, whose well-balanced, scientific mind valued truth above all else, it mattered not that the notions he wished to upset were hoary with age and hallowed by tradition. To put forward new ideas might evens suggest heresy, but Harvey pursued his experiments. For years he had been dissecting everything he could lay hand on – the bodies of men, animals, birds, frogs, snakes, rabbits, everything that might help him to solve his problem. The more he studied the more certain he became that his colleagues' ideas on blood movement were wrong.

The human heart consists of four chambers, the right and left auricles and the right and left ventricles. Dividing the heart vertically is the septum. In Harvey's time there had been no change in the theory of blood movement since Galen in the first century A.D. had made a few improvements on Aristotle. Fabricius and Sylvius, another sixteenth century anatomist, had made discoveries but the main body of opinion had remained substantially unchanged. This was that the blood originated from the liver and was of two different kinds. One kind, it was held came from the right ventricle of the heart and travelled through the body by way of the veins. The other came from the left ventricle and travelled to the body through the arteries. Both streams were believed to be slow and irregular and quite distinct.

Now Harvey had learnt from Fabricius that the veins had valves, and subsequent investigation confirmed this. These valves meant that blood could only flow through the veins in one direction, and that direction, Harvey found, was towards the heart. Therefore the idea that blood travelled through the veins from the right ventricle was erroneous.

Further, where was the blood coming from? He worked out the amount of blood that was entering the arteries. It was far too much for it to come from the stomach. The whole thing seemed unsatisfactory. Harvey continued his experiments, studied every available word that had been written on the subject and obtained a thorough knowledge of the anatomy of the body.

Then the solution came to him. There were not two different kinds of blood in the body. There was only one kind, the same in the veins as in the arteries. There was but one mass of blood that passed round and round, being driven on its journey by the heart, the body's motor. The blood was pumped from the heart, passed through the body in a " kind of circle," and returned once more to its source. The blood stream was a continuous circulation.

Not without the fullest possible research, not until he had viewed the matter from every angle, was Harvey satisfied that he had found the truth. Even then he did not rush into print. In 1616, twelve years before his book was published, he began to expound the theory in lectures delivered at the Royal College of Physicians. No one took much notice.

The volume was issued in 1628, and created a tremendous sensation in medical circles. Such a revolutionary theory could not be accepted without careful investigation. For a while Harvey's practice fell off. We are told that " 'twas believed by the vulgar that he was crackbrained and all the physicians were against him."

Yet in England opposition soon died down. After due inquiry the theory was accepted. Once it had been examined it was seen to be indisputably true. The doctor's practice not only recovered but grew considerably, for on his death he left a large fortune. On the Continent the theory met with more opposition. Much paper was wasted in attempts to refute it, and it was many years before due credit was given to it.

Harvey was now in high favour at Court. He was physician to Charles I, who displayed great interest in his work and gave him the use of his parks at Windsor and Hampton Court to aid him in his researches. Harvey had dedicated his book to the king, saying that the monarch was to his kingdom what the heart was to the body.

In 1636 Charles commanded him to accompany the Earl of Arundel on his embassy to the Emperor Ferdinand II. The man of science caused his noble colleague great anxiety by wandering off to look for specimens for dissection. Germany at that time was infested by robbers in consequence of the Thirty Years War, but it required more than thieves and vagabonds to prevent Harvey from searching for new specimens. He seems to have been disappointed by the poverty of material. "By the way," he wrote home, "we could scarce see a dog, crow, kite, raven or any bird, or anything to anatomise, only sum few miserable people, the reliques of the war and the plague, whom famine had made anatomies before I came."

Then came the Civil War. Harvey himself cared little for politics. He had one serious interest in life and that was medicine, but his sympathies were Royalist and he was the king's physician, so he left London with his master. Even while attending on his Majesty his thoughts were with his work. While the king's forces were gathering Harvey went to Derby to visit his friend, Percival Willoughby, in order to discuss with him the problem of uterine diseases.

He was present at the Battle of Edgehill. His immediate task was to take care of the Prince of Wales and the Duke of York, then lads of twelve and ten respectively. Harvey sought the shelter of a hedge not far from the fight and drew forth a book from his pocket. His study was rudely interrupted by a cannon-ball dropping nearby, and he thought it well to remove his royal charges to a safer spot.

Harvey was delighted when the king went to Oxford in the following year. He was then able to continue his work on anatomy, and incidentally was made a Doctor of Medicine of the university. It was during this period that his home in Whitehall was raided and searched. Invaluable manuscripts and anatomical preparations were taken away, and the loss was a serious one. His Royalist sympathies also nearly cost him his post at St. Bartholomew's, for the hospital raised the question of appointing a successor because Harvey "had withdrawn himself from his charge and retired to a party in arms against the Parliament."

At Oxford, Harvey did considerable research on the question of generation, and from the conclusions he reached there is little doubt that, given the necessary apparatus, he would have made startling and revolutionary discoveries. But he was in advance of his times. The use of a microscope was essential, and it had not yet been invented. Nevertheless, so important were the discoveries considered to be that when Harvey was persuaded by a friend to give him the manuscript for publication he said he felt like a second Jason with another Golden Fleece.

After Fairfax had captured Oxford the doctor returned to London to live in honourable retirement with his brothers, all of whom were prosperous merchants. He was sixty-eight years of age and much troubled with gout, which he sought to ease by putting his feet in cold water.

His theory of the circulation of the blood was now known and accepted everywhere. Even the European medical pundits had been forced to admit its truth. In 1654 the Royal College of Physicians wished to confer upon him the highest honour in the profession, that of president of the college. Harvey declined on account of his age, satisfying his desire to serve the institution by erecting a new building and equipping it with a well-filled library, a museum and a conversation room.

Harvey's busy life was now drawing to a close. His health was poor, though almost to the last he maintained his clarity of mind. On June 3, 1657, he was struck down by paralysis. Unable to speak, he managed to distribute some of his personal belongings to his nephews and then died. He was buried at Hempstead, in Essex (UK).

His wife having died childless some years previously, Harvey made a gift of his estate at Burmarsh, Kent, to the Royal College of Physicians, together with a fund for an annual lecture to be delivered at the college. In making this bequest Harvey urged the Fellows "to search out and study the secrets of nature by way of experiment, and also for the honour of the profession to continue mutual love and affection among themselves." The Harvey Oration is still delivered annually.

In 1883 the Fellows of the College had the remains of the great investigator removed to a white marble sarcophagus in the Harvey Chapel erected in Hempstead Church. In it they placed a copy of the large edition of his works.

Harvey's life was actuated by one great motive—to perfect mankind's knowledge of the human body so as to battle more successfully against disease and pain. His own discovery stands out as one of the greatest, made all the more glorious by the circumstances in which he worked. Without even the most elementary devices now used in research he was able to divine a truth that remains the greatest and most fundamental in the realm of physiology.

William H. Perkins

(1838–1907)

Great Chemist

People with colourful clothes and wearing an aroma of scent are common these days. But till the mid-nineteenth century only royal and affluent class could afford such trappings. In fact, the colour of clothes told the wearer's status. Today, there is a complete change in the scene. Even a common man in the street can easily afford these "privileges" of the bygone days. Modern synthetic dyestuff and perfume industries have brought about this revolutionary change adding colour and liveliness to every body's life. In the past, dyes and perfumes were expensive natural products extracted from rare plants and creatures, while they are today produced on a mass scale at a low price through sophisticated techniques. The chemist who made the invention of dye and wrought this revolution was William Henry Perkins.

Although Perkins' name had receded to the background due to the sparkling chemical inventions and discoveries made by chemists after him, his contributions to organic chemistry are no less than those of others. At the young age of 18, he made the chemical invention of aniline dye — or dye extracted from tar — the parent of half the dyes invented to date. At the age of 22, he planned, designed and founded the first modern dyestuff industry. No chemistry student can forget the valuable "Perkins reaction" for the synthesis of unsaturated acids. Apart from working on the magnetic rotatory power of substances, the formation of carbon rings, the chemistry of camphor, terpenes, berberine and other alkalis, he is also the inventor of another red dye called *"Alizarin"* and the first synthetic perfume *"Coumarin"*.

Perkins was born in 1838 to a successful builder and contractor of London. His father thought that if his son could design houses it would be of great help to him in his business. He therefore began to train young Perkins in architecture. Young Perkins did take up drawing and painting with zeal, but what he saw one day in the Royal College of Chemistry changed his career. Experiments on crystallization of substances then in progress at the college fascinated him so much so that he decided to become a chemist. His teacher Thomas Hall soon noticed his ardour for chemistry and took him to his famous German teacher August W. Hofmann. After talking to Perkins and observing his interest, Hofmann immediately understood that he had come across a genius. He offered him a job of research associate at the College of Chemistry. But Perkins' father had no faith in chemical experiments. He strongly opposed Perkins' desire to take up the job but eventually gave up when Hofmann personally visited Perkins' home and pursuaded him (his father) to allow his son to become a chemist. So, at the tender age of 15, Perkins was a researcher at the College of Chemistry. His interest in chemistry was so much aroused that he set up a small laboratory in his attic where he used to conduct research on problems of his interest after college hours.

Once, while listening to Hofmann's report on his experiments, Perkins came to know that quinine, an anti-malarial drug extracted from cinchona tree, could be prepared in laboratory. Hofmann himself had isolated a substance from tar which had composition almost similar to that of quinine. Intrigued, Perkins decided to try his hand at producing quinine in his laboratory at home. He selected the Easter vacations of 1856 to conduct his experiments. Through a series of tiresome experiments he found that it was possible to produce a substance which had composition almost similar to quinine but not exactly quinine. He was not able to determine that intermediary step which would convert that "near" quinine substance into quinine, the drug. During one of his experiments, he felt he had come very close to his goal. From the known chemical composition of quinine, he had simply to add oxygen to the substance — and he expected the formation of colourless quinine. He therefore took an oxidizing agent, potassium bichromate, and slowly poured it into the test-tube containing his "near" quinine substance. With high expectations he was looking at test-tube when to his dismay he found a deposit of a reddish brown powder at the bottom of the tube. He was annoyed. He threw the powder away and decided to repeat the experiment more carefully. Had there been any other chemist he would have given up this experiment and tackled the problem in a different manner. But Perkins was so sure of achieving his goal that he wanted to know where he had gone wrong. His careful experiments again fetched him that reddish black powder.

Out of curiosity, Perkins decided to examine the powder. He filtered it and found it black. In one of his subsequent experiments to examine it, he dissolved the powder in water. To his pleasent surprise, the water turned beautiful purple. Subsequently, what prompted Perkins to dip strips of clothes into the purple solution is not known because he knew quite little about dyeing. It is possible that some purple solution accidently spilled on his silk clothes. When the smudge could not be removed, the idea of a dye might have come to him and he might have pursued the trail. A dye had to be "fast" on a cloth so that it is not removed when the cloth is washed or it does not fade when exposed to sunlight. In other words, the dye has to be chemically linked to the cloth. In his new series of experiments with the black powder, Perkins found that it was fast on silk, but not on cotton or wool. Howsoever he washed dyed silk and howsoever long he exposed it to sunlight, the dye showed no change. He decided to produce it on a mass scale so that it could be used to dye silk clothes all over the country.

In his youthful enthusiasm, Perkins sent a sample of his dyed silk strips to Messrs Pullars of Perth, the leading manufacturer of dyestuff in those days. He had hoped that the manufacturer would immediately take up the production of his dye. After three anxious weeks, he received a letter from the manufacturer that the dye that he had invented was far superior than the best one then available. It could stand more rigorous tests than any other available dye. The letter however did not show any interest to manufacture the dye. Disappointed, Perkins nevertheless decided to patent the dye first before he would seek a manufacturer. Within a short time, he and his brother worked day and night and produced a few ounces of the dye and got it patented. Meanwhile, he also found tannin (used for tanning leather) to be the chemical, called "mordent" in chemistry parlance, which, when mixed with his dye, could become fast on cotton and wool. In other words, mordent acted as a link between cloth and the dye.

In due course, Perkins realised after talking with the manager of Messrs Pullars and another silk dye expert that it would not be as easy to produce his purple mauve on a large scale as he had thought. What looked so easy in his small laboratory was not so easy in a factory. It would take at least five to six years for manufacturing the dye on a large scale. Moreover, he found an inherent resistance against the new dye amongst the workers in the factories of aforementioned manufacturers because it meant changing their old ways of doing the work. Some workers even aired their doubts about the success of the dye. But Perkins, though hardly 20 years of age, was sure that his dye could be manufactured on a large scale quickly and would be widely accepted by the public. He knew that if his dye became a success he would not only become famous but also rich. And he wanted to become both fast.

Perkins had never been inside a chemical factory. Moreover, dyeing itself was a technique to which he was new. How to start a dye-making factory was therefore a challenge. But the very fact that he was a raw hand proved to be a blessing because he tackled all the problems in an entirely novel manner without any prejudices. In fact, had he looked for advice elsewhere he would not have got because synthetic dye-making itself was novelty. There was no other expert. Now that he decided to devote his time fully to his factory, he resigned his job at the College of Chemistry. It came as a shock to Hofmann for various reasons. One thing, Hofmann had contempt for commercial applications of chemistry. Secondly, he felt that Perkins' dye would not be a success. Lastly and more importantly, he felt chemistry was losing a genius. He wanted to see Perkins as one of the leading chemists of the time. But, fortunately, Perkins did not belie the expectations of Hofmann. In fact, had he continued research under Hofmann he may not have come in limelight as he did later. When Perkins decided to set up the factory, his father however gave him his full support because he was a businessman himself and wanted his son to possess worldly success. His brother too joined him. The three together raised a company "Perkins & Sons". In June 1857, Perkins' small factory began to function at Greenfield near Harrow. At every step he encountered problems because he was raising the factory from scratch. Nonetheless, it went on to become the first modern dyestuff industry in Britain and probably in the world.

Once Perkins' factory began to manufacture the dye on large scale, he came across yet another hurdle — the biggest of all in his desire to make his dye a commercial success. That hurdle was the dyers themselves, who were reluctant to change their old ways of dyeing a cloth. But here, too, Perking attained success by constant pursuasion and demonstration of his new dyeing techniques. He was thus not only a good scientist to invent a new dye and a good engineer to set up a factory but also a good salesman who could convince the dyers the advantages of his methods. In fact, had he not had one of these three qualities, he would have been a total failure and nobody would have known his name today. Once the reluctant dyers began to make use of

Perkins' dye, there was no looking back. The machine had been set in motion — and Perkins reaped the harvest. Shortly, purple mauve became a craze in the public. In those days, this particular colour was considered to be the preserve of only royal and rich persons because it was extracted from a rare and expensive plant. But Perkins' cheaper dye made possible for a common man to wear the colour. It caught the imagination of the public when in 1862 Queen Victoria, still mourning the death of Prince Consort, wore a mauve dress at an exhibiltion in Crystal Palace at Paris, France. The colour achieved so much popularity that even policemen used to call "you mauveon" to persons wearing clothes of mauve colour. Even the British Government brought out stamps of the same colour!

Along with commercial success, competitors too increased. To remain in the competition, Perkins not only made various shades of mauve but also invented other dyes which could be manufactured more cheaply than those extracted from plants and animals. He also invented a perfume "*Coumarin*" which was otherwise very expensive in those days. His key to success was to manufacture dyes and perfumes in his factory more cheaply than the natural, expensive ones available in the market. In due course, chemists all over the world woke up to this possibility of creating in their laboratory what was generally extracted from nature at a heavy price. News of the inventions of synthetic dyes began to pour in from all lands. In 1873, Germans entered the dyestuff industry in a big way, so much so that Perkins could not compete with them. At the age of 45, he sold his factory at a good price, built a good private laboratory and began to tackle problems of his own interest in organic chemistry. In 1906, King Edward VII knighted him for his contributions to dyestuff industry and declared him one of the greatest chemists of the time. The next year, on July 14, he passed away, leaving behind a proliferating chemical industry all over Europe.

William Herschel

(1738–1822)

Father of Star Astrology
The Discovery of Uranus

Nobody knows as yet who discovered the planets Mercury, Venus, Jupiter, Mars and Saturn. Perhaps, these have been known to man since the very dawn of human civilization. During the Vedic Age, the Aryans definitely knew about these five planets. The people of ancient times had also included the sun and the moon in the list of aforesaid planets because in the absence of telescopes no planet beyond these was visible to the unaided eyes.

William Herschel was born in England. In the beginning he was a music lover but from 1772 he started taking interest in astronomy and made many important astronomical discoveries till his last breath. He was also the most successful telescope-designer of his time. He was one of the founders of *"Star-astrology"*. He made an indepth study of planets and discovered many satellites. Towards the end of his life he was honoured with the Chairmanship of the "Royal Astronomical Society". Most of his studies were mainly related to stars. His discoveries of planets were only incidental. Nevertheless, Herschel will remain immortal in astronomy for advancing the knowledge of solar system beyond Saturn.

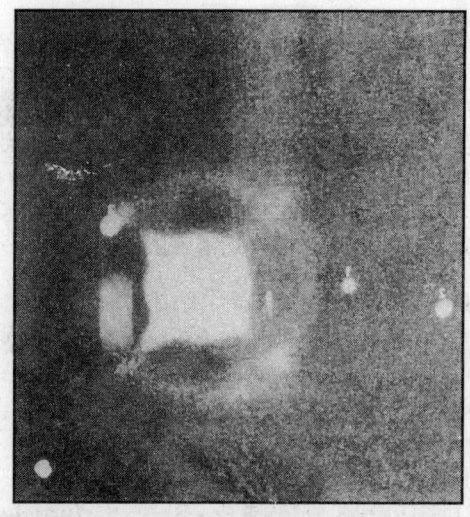

On March 13, 1781 when Herschel was making a telescopic observation of some small stars near the stars Castor and Pallux in the Gemini constellation, he saw a comet-like thing. With deeper observation it was found that that was not a comet but a planet. This is how Uranus was discovered. John Palmstead (1676-1719), the Royal Astronomer of England, had sighted this luminous object as many as six times during 1690-1715.

Various names were suggested for this new planet but ultimately "Uranus" became in vogue.

Physical Dimensions

The mean distance of Uranus from the sun is 2,868,600,000 kms. It revolves in its orbit around the sun at a speed of 4 miles per second and completes one revolution round sun in 84 years. Thus in one way it has been just a little over 5 years since Herschel discovered Uranus because a Uranus-year is equal to its 68,400 days. The equatorial diameter of Uranus is 51,800 kms. Thus in volume it is 64 times more than the size of the earth. Its density is much lesser than that of the earth, and hence, its mass is just about 15 times more than that of the earth.

According to Wilder, the diameter of its metallic core is 22,400 kms. It is covered by a 8,400 km thick layer of snow, which is covered with the atmosphere extending upto 4,700 km. Ramsay believed that Uranus is composed of water, methane and ammonia. Methane gas has been condensed as a result of the extreme cold conditions on Uranus. Hertzburg of Ottawa (Canada) has also found the existence of hydrogen on Uranus in free state. Possibly helium is also found there. Being an extremely cold planet, temperature at Uranus is almost below 0°C.

Uranus sometimes appears bright, sometimes dim. It is so far away from us that it is not possible to know anything about it without a telescope. If viewed through a small telescope only a small greenigh-yellow disc of Uranus is visible.

Satellites of Uranus

So far five satellites of Uranus have been discovered namely Miranda, Ariel, Umbriel, Titania and Oberon. Herschel was under the wrong impression that he had discovered six satellites of Uranus because four out of those were, in fact, stars. The remaining satellites of Uranus were Titania and Oberon. Two more satellites, viz., Ariel and Umbriel were discovered by a British astronomer, William Lassell, in 1851. The fifth satellite Miranda was discovered by Gerard Kupier in 1948. When this planet's pole faces the earth, all satellites are seen revolving in circular orbits. When its line of Equator is turned towards the earth, the satellites are seen revolving like vertical discs. The average distance of Miranda from Uranus is 1,21,600 kms, and that of Oberon is 5,82,400 kms.

Uranus is a strange planet. On this planet's soil the sun rises in the west and sets in the east.

No final data are available about the physical dimensions of Uranus's satellites. The information gathered so far suggests that the diameter of Miranda, Umbriel, Ariel, Oberon and Titania is about 550 kms, 1,000 kms, 1,500 kms, 950 kms and 1,800 kms, respectively.

William Shockley

(1910–1989)

Transistor? On, you mean, my pocket radio? Any body will ask as soon as he or she hears the word "Transistor". But, then, G. Marconi is the inventor of radio — and how come William

Shockley and his team invented transistor? he or she may further ask. Indeed, it is not his or her fault for taking a pocket radio as transistor because the term has been loosely used. In reality, a transistor is simply one of the crucial parts of a pocket radio. Today, it is in fact a crucial part of almost all the electronic gadgetry that is seen in the world. From homes, offices, industries to computers, satellites and spacecraft, it is used almost everywhere. The invention of transistor has created a revolution unsurpassed in the history of civilization. And mind it, it is simply a chip of a material with of course miraculous electrical properties. Its inventor was not a single individual with an unclear goal. It was a team of scientists led by William Shockley that invented transistor which was a planned goal. This story of the invention is therefore different from others in several respects because it is difficult to follow how each scientist made his contribution to the invention. Nevertheless, the key role behind the invention of transistor was of William Shockley. It is better to consider his life history and introduce other scientists as and when required.

William Shockley was born on February 13, 1910 at London, England, to American parents. His father was a mining engineer and his mother a mineral surveyor. When he was three, the family shifted to Palo Alto, California, U.S.A., a place today renowned as the Silicon Valley. Both of his parents were of the opinion that school education was not proper for children and did not send young Shockley to school till the age of eight. His father took keen interest in his studies and always discussed scientific subjects with him. His mother also taught him mathematics. Although young Shockley went to school, he himself never liked it. In those early years, the biggest impression on his mind was created by his neighbourer, Professor P.R. Ross who taught physics at Stanford University. Young Shockley used to visit Professor Ross's house to play with his daughters of his age. Professor Ross knew how to talk with children and engage their attention. He became friendly with young Shockley and stimulated his interest in physics. When he used to read the biographies of giants of science such as Isaac Newton, Albert Einstein and others, who had made valuable contributions to physics at a very early age, he used to wonder whether he himself would ever be able to do so. He always strived to score more marks than his fellow students. Besides physics, he had a great interest in magic. Occasionally, he used to give amateur shows before his neighbourers.

During his graduation at California Institute of Technology, Robert Millikan, the Nobel Laureate, who measured the electric charge on a single electron, a negatively charged particle, was its head. Young Shockley had high regards for Millikan as a scientist but he was surprised to find that he (Millikan) was not at all engaged in any research at that later age of his life. However, it was only in course of time that he realised that it was important work that Millikan was performing at the institute. He realised that it was important not only to do excellent scientific research but also to encourage young scientists in their work and provide them the facilities that they require. This realisation which came so early in his life helped him later in organising research which led to the invention of transistor, a product of the efforts of several scientists, young and old. In 1936 after doing Ph.D at Massachusetts Institute of Technology on the electrical properties of crystals, Shockley joined the Bell Telephone Laboratories on the persuasion of Mervin J. Kelly, the then research director of the laboratory.

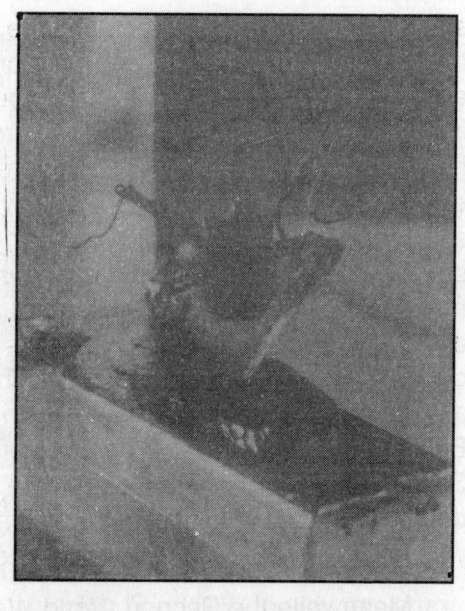

Basically, Shockley wanted to work under the guidance of the charming Nobel Laureate C.J. Davisson who had experimentally shown that an electron behaves like a wave as well as a particle. But Kelly had some different ideas. He appointed him as trainee in his own department of vacuum tubes, an "in"

subject in those days. A vacuum tube, as the name implies, is a tube containing almost vacuum through which the flow of electrons — electric current — could be controlled. In those days, it was a common feature of all electronic gadgets and, though having several defects, was considered beyond perfection. One of the favourite schemes of Kelly was to improve telephone exchanges. The telephone exchanges of those days were composed of mechanical parts which needed repair and maintenance regularly. Kelly thought that the mechanical parts could be replaced by vacuum tubes or some such electronic devices. He used to talk at length on this subject to Shockley always provoking him to work along these lines because after all Bell Laboratories was devoted to telephone equipment. Certainly, Shockley began to think along these lines. In course of time it became his dream to build an electronic telephone exchange. It was his first step in the direction of the invention of transistor.

Meanwhile the Second World War had started and took in its fold Shockley like other young scientists. For the next two years he therefore pursued projects of interest to the war. After the war when he returned to Bell Laboratories, he was made the co-project director. He took the subject of the electrical behaviour of crystals again, which he had left after his Ph.D. Although crystals were then not popular due to the widespread use of vacuum tubes, the Bell Laboratories allowed him to pursue his interests. As a co-project director, Shockley had to lead a team of scientists, each an expert in his own field. Among a few other in the team John Brattain was an adept experimentalist. Later, John Bardeen, a theoretical physicist, joined the team accidently because of dearth of room elsewhere in the laboratory. Ideally speaking, Shockley, Brattain and Bardeen formed the best team for the invention of transistor. Shockley had the dream and will to invent transistor, Bardeen had a sound theoretical understanding of crystals to project the dream on to paper in black and white, and Brattain could convert the ideas on paper into real, concrete things.

The basic aim with which Shockley and his team began research was to invent an amplifier, an electrical device which could amplify electric current. The device was to be built of crystal. In those days, vacuum tube was used to amplify an electric signal but Shockley and his team wanted to try a crystal instead to achieve the same. He believed that if a crystal could detect radio waves like a vacuum tube, it could as well amplify an electric signal or current. This crystal material is technically known as "semi-conductor" because its capacity to conduct electric current is semi — between that of good conductor of electric current such as metal and that of bad conductor such as wood. Some substance, often called "impurities", are added to the semi-conductor material to impart new electrical properties to it; they especially affect the flow of electric current or electrons through it. In the beginning, Shockley thought that the semiconductor such as silicon or germanium would amplify an electric signal if some external electric field could influence it. But howsoever the team tried, all efforts came to a naught. Shockley called these failures "creative failures" because they enabled the team to understand semiconductors and they were the stepping stones to success.

When Shockley and his team began their research on semiconductors, it was also the most opportune time. Quantum physics, which gives a picture of what is happening inside a semi-conductor or for that matter any material, was already developed. On the other hand, technology needed to produce semiconductors of desired properties had also advanced. Both these developments enabled Shockley and his team to understand the internal structure of semiconductors in the first place and, second, to produce a semiconductor of desired interest. Nevertheless, there were several trials and errors and the team pursued their goal. On the basis of his understanding of semiconductors, Bardeen suggested that heating or cooling the crystal would produce the desired effect but all proved to be of no avail. Then began a series of experiments in which the semiconductor was kept in an insulating liquid and then conducting liquid. When the semiconductor was kept in a conducting liquid some changes in its electrical properties were observed. In due course, the conducting liquid was replaced by a drop of it and eventually by a metal contact. A gold leaf cut into two by a razor blade so that these two contact points were hardly 2000th of an inch apart on a piece of germanium which was kept on a metal plate, the third contact. This was the first transistor which amplified electric signals fed to it several times over. The day was December 16, 1947, a few days before the Christmas. The amplification phenomenon later came to be known as the "Transistor effect".

Almost the next day Shockley called other scientists and officials of the Bell Laboratories for the demonstration of his team's crystal amplifier. Before the small gathering, the famous words of Alexander Bell, the founder of the Bell Laboratories, "Mr Watson, come here...", which he had said after he had accidentally invented telephone were amplified. A loudspeaker blared out the words. In a way, the invention was a step towards fulfilment of Bell's ideals to remove deafness. At that time everybody in the meeting, including the three inventors, knew the importance of the invention but could not visualise how much it would revolutionise electronics as it has since. For the next five months, the invention remained a secret. It was improved upon and the patents drawn. Even when it was publicly announced on June 30, 1948, it did not cause any ripple. Even the *New York Times* gave hardly 36 lines on it at the end of their "News of Radio" column. Although the successful creation of a semiconductor amplifier was talked about in the scientific circles, the word "transistor" was nowhere either in print or heard about. In fact, that chip of semiconductor which amplified electric signals had no name for several months after its invention. Eventually, the Bell Laboratories organised an in-house competition for the name. Among the several names suggested such as amplister, transistor, etc, the name "transistor" was selected. It was suggested by a physicist John K.Pierce because semiconductor was a resistor which allowed the transfer of electric current through it. In view of the immediate acceptance of transistor in electronics industry, its revolutionary effects in all walks of life and its use in understanding semiconductors, Shockley, Bardeen and Brattain were jointly awarded the 1956 Nobel Prize for physics.

What is transistor and why is it so important as to cause a revolution in electronics industry and elsewhere? A transistor is like a water tap. Any ordinary tube, like any ordinary conductor of electric current allows water to flow but it is the tap which controls the flow. By turning the tap, water could be allowed to drip in droplets or flow in a torrent. It could also be totally stopped. Moreover, a slight turning of a tap could drastically change the amount of water flow. In a transistor there are three contact points connected to three wires, of which two become a part of the electric circuit like the tube allowing water to flow through it. The third wire acts like the control of the tap. The transistor can therefore behave like an electric switch to put "on" or "off" electric current flow: an amplifier to step up or step down the flow of electric current; a regulator of flow of electric current. In short, it could perform almost all the functions of vacuum tube with none of its defects. For instance, a transistor cannot break down as a vacuum tube does; it needs low power to function; it does not need time to warm as a vacuum tube needs; it is far smaller in size than a vacuum tube, so on. In other words, a transistor not only replaced a vacuum tube which was then commonly used in all electronic gadgetry but was also more reliable, cheaper and smaller in size. It visibly spurred on a revolution towards miniaturisation of electronic components and goods. Without the invention of transistor certain things would have been difficult to attain and certain things impossible. For instance, using vacuum tubes, a computer was of the size of a big hall and needed a huge airconditioning systen to cool down the hot tubes. Now a computer is of the size of a palm. Thanks to transistor, it is now possible to have small hearing aids which could even be incorporated in a spectacle, pacemakers which could be incorporated in a heart to regulate its beating, so on. In fact, without minielectronic components, Space Age would not have been possible. And, of course, transistor made possible an electronic telephone exchange, the dream with which Shockley began research.

Today, Brattain and Shockly are no more alive, but Bardeen continues to pursue their research interests. Bardeen, in fact, went on to win another Nobel Prize for physics for his researches in superconductivity. On the other hand, Shockley continued his researches on semiconductors. In 1954, he gave up research at Bell Laboratories and started his own company Shockley Semiconductor Laboratories at Palo Alto, California, the place where he had spent his boyhood. Several other companies also followed him to cash in on the transistor boom. However, Shockley was not successful and he returned to pursue his research at Stanford University. In the meanwhile, Palo Alto came to be known as the Silicon Valley because Silicon is the semiconductor material commonly used there. Shockley however continued to guide and teach students until his death on August 12, 1989. He believed that his success at inventing transistor was due to what he called "*creative failures*", "*the will to think*" and "*Try simple cases.*"

William Thomson, Sir (Lord Kelvin)

(1824–1907)

Great Physicist

"I am never content until I have constructed a mechanical model of the studying.

If I succeed in making one, I understand; otherwise I do not."

About 167 years ago, a boy of ten took his matriculation (school final) examination with young men seven and eight years older to him at Glasgow University in Scotland. The boy's name was William and his father, James Thomson, was professor of mathematics at the university.

Born in Belfast, Ireland, in 1824, the British physicist began the papers on the laws of conservation and dissipation of energy.

To pass the matriculation examination at the age of ten was indeed an extraordinary achievement. His sister records that before William was ten years old he would try to solve some of the problems his father had set his class at the university!

When he was about seventeen, he joined the Peterhouse College at Cambridge. After completing Cambridge, he went to Paris where he studied under the eminent scientist, Professor

Regnault, who was at that time engaged in his research on the thermal properties of steam.

Association with Professor Regnault inspired him to take up research in physics. His knowledge in physics was so great that he was appointed as professor of Natural Philosophy (science) at Glasgow University at the young age of twenty-two. He held this post for fifty-three years, and during this time, became one of the greatest physicists of the world.

One of his early researches was about the age of the earth. The theory about the age of the earth current at that time supposed that the earth existed very much in the same form for some thousands of millions of years. Thomson showed with his experiments and calculations on heat and the dissipation of energy, that this period was about one hundred to five hundred million years only.

In 1851 he formulated his famous *Law of Conservation* which states that energy is not lost, but gets transformed to other forms of energy. This means that the total energy in a closed system is always the same (constant), that is, energy is conserved in any transfer from one form to another.

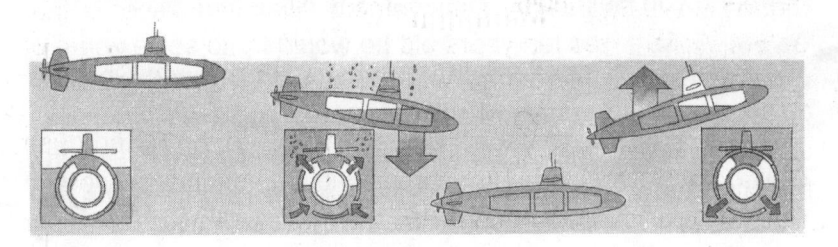

There are many inventions to his credit. The submarine cable is one. The Mirror Galvanometer is another. Anything that would lighten labour or help science interested Thomson. The first electric meter was his creation. He invented instruments that could measure tiny currents of one thousandth of an ampere to ten thousand amperes and electric pressures from minute fractions of a volt to one hundred thousand volts. He held patent rights over seventy inventions. These inventions were related mainly to telegraphy, electrical measurements, and marine navigation. He had published papers and other works on every important area of 19th century physics. These publications number more than 661.

In spite of his great achievements and vast knowledge, William Thomson was a very simple and kind-hearted man. He praised and encouraged his students and helped them with their studies and researches.

In 1852 he married Margaret, the daughter of his father's great friend, Walter Crums. They were a couple devoted to each other. Thomson loved to travel and Margaret was his constant companion, whether on land or sea. The one sorrow in their life was that they had no children. Also Margaret's health was delicate. So, under medical advice, Thomson used to take Margaret abroad frequently, and it was in one of those trips that he met the great German Physicist Hermann von Helmholtz. They became close friends.

In spite of all his efforts, Margaret's health declined. She was an accomplished woman who wrote poems, and even in sickness she continued to write. A friend wrote of her that she was sincere, bright and very keen- minded, and with a true insight into the real worth of things and thoughts. She died in 1870 and Thomson was heart-broken.

To forget the pain, Thomson travelled ceaselessly. Four years later, during one of these travels he met Frances Blandy, who succeeded in taking him out of himself. They were married soon, and his second marriage turned out to be as happy as his first one.

He retired from Glasgow University in 1899. Now he seldom attended public functions but *followed* every development in science with interest. Even though he was seventy-six *when* he left his professorship, he didn't like to l*eave the* university. At the beginning of the next academic year, an old man walked into the registration room *alongwith a* crowd of undergraduates and signed on the register 'Lord Kelvin – research student'.

Now the health of Frances began to worry him. In 1907, she had a stroke which partly paralysed her. He himself caught a chill, which developed into a serious sickness. It became worse and on Dec. 17, 1907, he passed away. He was buried at Westminster Abbey next to the grave of Sir Issac Newton.